Advances in Multifunctional Materials and Systems

Advances in Multifunctional Materials and Systems

Ceramic Transactions, Volume 216

A Collection of Papers Presented at the 8th Pacific Rim Conference on Ceramic and Glass Technology
May 31–June 5, 2009
Vancouver, British Columbia

Edited by
Jun Akedo
Hitoshi Ohsato
Takeshi Shimada

Volume Editor
Mrityunjay Singh

The
American
Ceramic
Society

A John Wiley & Sons, Inc., Publication

Published by John Wiley & Sons, Inc., Hoboken, New Jersey.
Published simultaneously in Canada.

No part of this publication may be reproduced, stored in a retrieval system, or transmitted in any form or by any means, electronic, mechanical, photocopying, recording, scanning, or otherwise, except as permitted under Section 107 or 108 of the 1976 United States Copyright Act, without either the prior written permission of the Publisher, or authorization through payment of the appropriate per-copy fee to the Copyright Clearance Center, Inc., 222 Rosewood Drive, Danvers, MA 01923, (978) 750-8400, fax (978) 750-4470, or on the web at www.copyright.com. Requests to the Publisher for permission should be addressed to the Permissions Department, John Wiley & Sons, Inc., 111 River Street, Hoboken, NJ 07030, (201) 748-6011, fax (201) 748-6008, or online at http://www.wiley.com/go/permission.

Limit of Liability/Disclaimer of Warranty: While the publisher and author have used their best efforts in preparing this book, they make no representations or warranties with respect to the accuracy or completeness of the contents of this book and specifically disclaim any implied warranties of merchantability or fitness for a particular purpose. No warranty may be created or extended by sales representatives or written sales materials. The advice and strategies contained herein may not be suitable for your situation. You should consult with a professional where appropriate. Neither the publisher nor author shall be liable for any loss of profit or any other commercial damages, including but not limited to special, incidental, consequential, or other damages.

For general information on our other products and services or for technical support, please contact our Customer Care Department within the United States at (800) 762-2974, outside the United States at (317) 572-3993 or fax (317) 572-4002.

Wiley also publishes its books in a variety of electronic formats. Some content that appears in print may not be available in electronic format. For information about Wiley products, visit our web site at www.wiley.com.

Library of Congress Cataloging-in-Publication Data is available.

ISBN 978-0-470-89058-5

Printed in the United States of America.

10 9 8 7 6 5 4 3 2 1

Contents

MICROWAVE MATERIALS

Preface

The symposia *Advances in Electroceramics* and *Microwave Materials and Their Applications* were held during the 8th Pacific Rim Conference on Ceramic and Glass Technology (PACRIM 8) from May 31–June 5, 2009 in Vancouver, Canada. This issue contains 17 peer-reviewed papers (invited and contributed) from these two symposia.

We would like to thank all members of the organizing committee for their work in the planning and execution of these symposia. The editors wish to extend their gratitude and appreciation to all the authors for their contributions, to all the participants and session chairs for their time and effort, and to all the reviewers for their valuable comments and suggestions. The invaluable assistance of the staff of the meeting and publication departments of The American Ceramic Society is gratefully acknowledged.

We hope that the collection of papers presented here will serve as a useful reference for the researchers and technologists working in the field of electronic ceramic materials and devices.

JUN AKEDO, *National Institute of Advanced Industrial Science and Technology, Japan*
HITOSHI OHSATO, *Nagoya Institute of Technology, Japan*
TAKESHI SHIMADA, *Hatachi Metals, LTD, Japan*

Introduction

The 8th Pacific Rim Conference on Ceramic and Glass Technology (PACRIM 8), was the eighth in a series of international conferences that provided a forum for presentations and information exchange on the latest emerging ceramic and glass technologies. The conference series began in 1993 and has been organized in USA, Korea, Japan, China, and Canada. PACRIM 8 was held in Vancouver, British Columbia, Canada, May 31–June 5, 2009 and was organized and sponsored by The American Ceramic Society. Over the years, PACRIM conferences have established a strong reputation for the state-of-the-art presentations and information exchange on the latest emerging ceramic and glass technologies. They have facilitated global dialogue and discussion with leading world experts.

The technical program of PACRIM 8 covered wide ranging topics and identified global challenges and opportunities for various ceramic technologies. The goal of the program was also to generate important discussion on where the particular field is heading on a global scale. It provided a forum for knowledge sharing and to make new contacts with peers from different continents.

The program also consisted of meetings of the International Commission on Glass (ICG), and the Glass and Optical Materials and Basic Science divisions of The American Ceramic Society. In addition, the International Fulrath Symposium on the role of new ceramic technologies for sustainable society was also held. The technical program consisted of more than 900 presentations from 41 different countries. A selected group of peer reviewed papers have been compiled into seven volumes of The American Ceramic Society's Ceramic Transactions series (Volumes 212-218) as outlined below:

- **Innovative Processing and Manufacturing of Advanced Ceramics and Composites, Ceramic Transactions, Vol. 212,** Zuhair Munir, Tatsuki Ohji, and Yuji Hotta, Editors; Mrityunjay Singh, Volume Editor
  Topics in this volume include Synthesis and Processing by the Spark Plasma

Method; Novel, Green, and Strategic Processing; and Advanced Powder Processing

- **Advances in Polymer Derived Ceramics and Composites, Ceramic Transactions, Vol. 213,** Paolo Colombo and Rishi Raj, Editors; Mrityunjay Singh, Volume Editor
This volume includes papers on polymer derived fibers, composites, functionally graded materials, coatings, nanowires, porous components, membranes, and more.

- **Nanostructured Materials and Systems, Ceramic Transactions, Vol. 214,** Sanjay Mathur and Hao Shen, Editors; Mrityunjay Singh, Volume Editor
Includes papers on the latest developments related to synthesis, processing and manufacturing technologies of nanoscale materials and systems including one-dimensional nanostructures, nanoparticle-based composites, electrospinning of nanofibers, functional thin films, ceramic membranes, bioactive materials and self-assembled functional nanostructures and nanodevices.

- **Design, Development, and Applications of Engineering Ceramics and Composite Systems, Ceramic Transactions, Vol. 215,** Dileep Singh, Dongming Zhu, and Yanchun Zhou; Mrityunjay Singh, Volume Editor
Includes papers on design, processing and application of a wide variety of materials ranging from SiC SiAlON, ZrO_2, fiber reinforced composites; thermal/environmental barrier coatings; functionally gradient materials; and geopolymers.

- **Advances in Multifunctional Materials and Systems, Ceramic Transactions, Vol. 216,** Jun Akedo, Hitoshi Ohsato, and Takeshi Shimada, Editors; Mrityunjay Singh, Volume Editor
Topics dealing with advanced electroceramics including multilayer capacitors; ferroelectric memory devices; ferrite circulators and isolators; varistors; piezoelectrics; and microwave dielectrics are included.

- **Ceramics for Environmental and Energy Systems, Ceramic Transactions, Vol. 217,** Aldo Boccaccini, James Marra, Fatih Dogan, Hua-Tay Lin, and Toshiya Watanabe, Editors; Mrityunjay Singh, Volume Editor
This volume includes selected papers from four symposia: Glasses and Ceramics for Nuclear and Hazardous Waste Treatment; Solid Oxide Fuel Cells and Hydrogen Technology; Ceramics for Electric Energy Generation, Storage, and Distribution; and Photocatalytic Materials.

- **Advances in Bioceramics and Biotechnologies, Ceramic Transactions, Vol. 218;** Roger Narayan and Joanna McKittrick, Editors; Mrityunjay Singh, Volume Editor
Includes selected papers from two cutting edge symposia: Nano-Biotechnology and Ceramics in Biomedical Applications and Advances in Biomineralized Ceramics, Bioceramics, and Bioinspiried Designs.

I would like to express my sincere thanks to Greg Geiger, Technical Content Manager of The American Ceramic Society for his hard work and tireless efforts in the publication of this series. I would also like to thank all the contributors, editors, and reviewers for their efforts.

MRITYUNJAY SINGH
Volume Editor and Chairman, PACRIM-8
Ohio Aerospace Institute
Cleveland, OH (USA)

Electroceramics

NANOSTRUCTURED CERAMICS
OF PEROVSKITE MORPHOTROPIC PHASE BOUNDARY MATERIALS

M. Algueró, H. Amorín, T. Hungría, J. Ricote, R. Jiménez and A. Castro
Instituto de Ciencia de Materiales de Madrid (ICMM), CSIC. Cantoblanco, 28049 Madrid, Spain

P. Ramos
Departamento de Electrónica, Universidad de Alcalá. 28871 Alcalá de Henares, Spain

J. Galy
Centre d'Elaboration de Matériaux et d'Etudes Structurales (CEMES), CNRS. 29 rue Jeanne Marvig, BP 94347, 31055 Toulouse, France

J. Holc and M. Kosec
Institute Jozef Stefan. Jamova 39, 1000 Ljubjlana, Slovenia

ABSTRACT
    In this paper, we review and update current research at ICMM on the processing of submicron- and nanostructured ceramics of high sensitivity piezoelectric materials, and on the study of the grain size effects on the macroscopic properties. Nanocrystalline powders synthesised by mechanochemical activation are being used for the processing. Firstly, ceramics of $(1-x)PbMg_{1/3}Nb_{2/3}O_3-xPbTiO_3$ in the MPB region (x=0.2 and 0.35) with grain size in the submicron range down to the nanoscale (~90 nm), processed by hot pressing of the nanocrystalline powder are introduced. The study of the electrical properties as a function of grain size indicated an evolution from ferroelectric to relaxor type behaviour with the decrease in size, which is proposed to result of the slowing down of the kinetics of the relaxor to ferroelectric transition in the submicron range, and of the stabilisation of intermediate domain configurations. Nevertheless, significant piezoelectric activity can be obtained after poling under tailored conditions that speed up the kinetics. Secondly and deeper in the nanoscale, MPB $PbZn_{1/3}Nb_{2/3}O_3-PbTiO_3$ and $PbZn_{1/3}Nb_{2/3}O_3-PbFe_{1/2}Nb_{1/2}O_3-PbTiO_3$ ceramics with a grain size of 15-20 nm, processed by spark plasma sintering of the nanocrystalline powder are presented. At these very small grain sizes, the high temperature relaxor state is definitively stabilised, which dynamics presents distinctive features. Finally, the same experimental approach has been applied to MPB $BiScO_3-PbTiO_3$ and materials with an average grain size from 80 to 20 nm have been processed. Macroscopic ferroelectricity and piezoelectricity have been demonstrated for the largest sizes.

INTRODUCTION
    Piezoelectric devices such as multilayer actuators are not oblivious to the current miniaturisation trends in ceramic technologies for microelectronics. The thickness of the ceramic layers is being reduced down to the micron, which requires the decrease of the grain size down to the submicron range close to the nanoscale for reliability.[1] Also, ferroelectric polycrystalline films are integrated in microelectromechanical systems to implement sensing and actuation.[2] The feasibility of these technologies depends on how the macroscopic functional properties scale with grain size. Size effects have been extensively studied for the electrical properties; basically permittivity and ferroelectric hysteresis loops, of ceramics of perovskite tetragonal $BaTiO_3$ down to a grain size of 30 nm.[3] However, reports on the electromechanical properties of submicron- and nanostructured ceramics of highly piezoelectric materials are scarce. A significant decrease of the piezoelectric coefficients of

soft PZT with grain size has been reported in the submicron range (down to 0.17 μm), which was associated with the clamping of the ferroelectric/ferroelastic domain walls.[4]

We have focused on alternative perovskite morphotropic phase boundary (MPB) materials that show higher single crystal piezoelectric coefficients (less wall contribution) than $Pb(Zr,Ti)O_3$. On the one hand, we addressed relaxor based MPB systems, such as $PbMg_{1/3}Nb_{2/3}O_3$-$PbTiO_3$ and $PbZn_{1/3}Nb_{2/3}O_3$-$PbTiO_3$. Single crystals of these materials present ultrahigh piezoelectricity with d_{33} coefficients above 2000 pC N^{-1}.[5] On the other hand, we are also working on MPB $BiScO_3$-$PbTiO_3$, a material with a phase diagram analogous to that of $Pb(Zr,Ti)O_3$ in the MPB region, and higher piezoelectric coefficient and Curie temperature.[6] Here, we review and update our research on the processing of submicron and nanostructured ceramics of these systems, and on the study of their macroscopic properties as a function of grain size.

SUBMICRON- AND NANOSTRUCTURED CERAMICS OF MPB $PbMg_{1/3}Nb_{2/3}O_3$-$PbTiO_3$

Nanocrystalline Powder

Nanocrystalline powders of $0.65PbMg_{1/3}Nb_{2/3}O_3$-$0.35PbTiO_3$ were synthesised by mechanochemical activation of PbO (99.9+%, Aldrich), MgO (98%, Aldrich), TiO_2 (99.8%, Alfa Aesar), and Nb_2O_5 (99.9%, Aldrich) without any excess of MgO or PbO. The activation was carried out with a high energy planetary mill (Model PM 400, Retsch) and tungsten carbide milling media. Details of the procedure and of the mechanisms taking place during the process can be found elsewhere.[7] The final powder consisted of a major nanocrystalline perovskite phase (~ 25 nm crystal size), a ~30 wt% of amorphous phase depleted of MgO, and trace amounts of pyrochlore and crystalline MgO. Contamination levels were below 50 and 600 ppm for Co and W, respectively. Powder was lastly de-agglomerated by attrition milling in isopropanol, which resulted in particles with a size distribution of 10, 50 and 90 wt% under 0.37, 0.84 and 5.49 μm, respectively.

Processing by Hot Pressing

Nanocristalline powders of this type have been used to process coarse grained ceramics of (1-x)$PbMg_{1/3}Nb_{2/3}O_3$-$xPbTiO_3$ with x=0.2, 0.3, 0.35 and 0.4 with high chemical homogeneity and crystallographic quality by sintering in PbO. Also, the processing of submicron grain size ceramics by hot pressing of the powder was demostrated.[8] The feasibility of obtaining dense ceramics with decreasing grain size in the submicron range down to the nanoscale was systematically studied for x=0.2. The lowest temperature that could be used was 700ºC, below which a non-negligible amount of pyrochlore phase was stabilised. This second phase resulted of the crystallisation of the amorphous fraction, and a temperature above 700ºC was necessary to fully transform it into perovskite. Nevertheless, basically perovskite single phase ceramics were obtained by hot pressing between 1000 and 700ºC, which presented a decreasing average grain size from 0.25 μm down to 90 nm and densifications above 90%.[9]

The same approach has been used for x=0.35. Nanocrystalline powders were hot pressed with 60 MPa at temperatures between 1000 and 700ºC (± 3ºC min^{-1} heating and cooling rates, 1 h soaking time, pressure applied at the target temperature). X-ray diffraction (XRD) patterns for ceramics of (1-x)$PbMg_{1/3}Nb_{2/3}O_3$-$xPbTiO_3$ with x=0.35, hot pressed at 800 and 700ºC are shown in Figure 1 along with patterns for the materials with x=0.2 for comparison (Cu K_α radiation, D500 diffractometer, Siemens). Note that results for x=0.35 were analogous to those for x=0.2, and perovskite single phase materials were obtained at 800ºC, while traces of pyrochlore phase still persisted after hot pressing at 700ºC. Densifications above 95% were obtained in this case.

Also, similar grain sizes were obtained as it is illustrated in Figure 2, where scanning force microscopy (SFM) images for ceramics with x=0.35 and 0.2 are shown (Nanotec Electrónica).

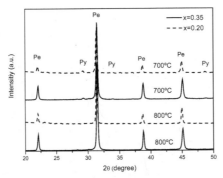

Figure 1. XRD patterns of ceramics of $(1-x)PbMg_{1/3}Nb_{2/3}O_3$-$xPbTiO_3$ with x=0.35 and 0.2, processed by hot pressing of nanocrystalline powder synthesised by mechanochemical activation. Pe: perovskite, Py: pyrochlore.

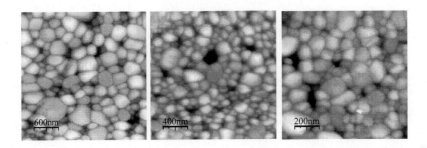

Figure 2. SFM images of ceramics of $(1-x)PbMg_{1/3}Nb_{2/3}O_3$-$xPbTiO_3$ processed by hot pressing of nanocrystalline powder synthesised by mechanochemical activation. Left: x=0.35 HP at 900°C, centre: x=0.2 HP at 800°C, right: x=0.2 HP at 700°C.

Electrical and Electromechanical Properties

$0.65PbMg_{1/3}Nb_{2/3}O_3$-$0.35PbTiO_3$ is at the core of the MPB region, and coarse grained ceramics processed from the nanocrystalline powder present a d_{33} piezoelectric coefficient of 525 pC N^{-1} after poling at 3 kV mm^{-1} (field applied at 150°C and maintained during cooling down to room temperature). The electrical properties of materials hot pressed at 900°C, which had an average grain size of 0.15 µm have been already reported. These submicron- structured ceramics showed relaxor type electrical behaviour. Nevertheless, Rietveld analysis of XRD data clearly indicated that the material was in the monoclinic *Pm* phase at room temperature (RT), and ferroelectricity was further confirmed by poling that resulted in a d_{33} of 290 pC N^{-1}.[10]

We show here results for materials hot pressed at 800 and 700°C, for which the average grain size has already entered the nanoscale.

The temperature and frequency dependences of the permittivity are shown in Figure 3 for the two materials, along with those for a ceramic hot pressed at 900°C (dynamically measured during heating at 1.5°C min^{-1}, HP4284A precision LCR meter, Agilent Technologies).

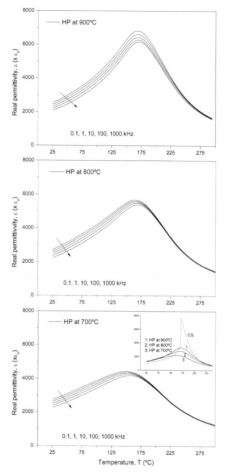

Figure 3. Temperature and frequency dependences of the real permittivity
of $0.65PbMg_{1/3}Nb_{2/3}O_3$-$0.35PbTiO_3$ ceramics processed by hot pressing of the nanocrystalline powder
at decreasing temperatures. The inset shown the permittivity at 1 KHz for the different materials
compared with that of a coarse grained ceramic (dashed line)

Ferroelectric hysteresis loops for the materials hot pressed at decreasing temperatures are shown
in Figure 4, along with that for a coarse grained ceramic for comparison (high voltage sine waves were
applied by a combination of a synthesiser/function generator, HP 3325B, Agilent, and a high voltage
amplifier, model 10/40A, Trek, and loops were measured with a home-built charge to voltage converter
and software for data acquisition and analysis, ICMM, CSIC).

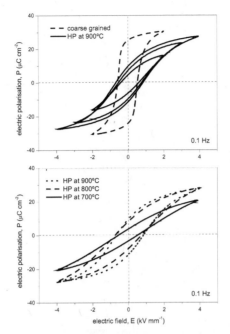

Figure 4. Room temperature ferroelectric hysteresis loops of $0.65PbMg_{1/3}Nb_{2/3}O_3$-$0.35PbTiO_3$ ceramics processed by hot pressing of the nanocrystalline powder at decreasing temperatures. The loop for a coarse grained material is also shown for comparison.

All submicron and nanostructured ceramics presented dependences of permittivity on frequency and temperature typical of relaxors, unlike coarse grained materials that show a first order like ferroelectric to relaxor transition (see inset in Fig. 3). The main effect on decreasing grain size is the reduction of the height of the maxiumum in permittivity, and its slight shift towards lower temperatures. Most of the polarisability of these materials, and the dispersive maximum in permittivity, are associated with the dynamics of the polar nanoregions (PNRs) characteristic of the relaxor state. The grain size effect thus, is related to the disturbance of this dynamics at the grain boudaries, an effect previously discussed in depth for the material with 0.15μm grain size.[11] The possibility of a non polar, highly defective grain boundary seems unlikely, for Maxwell-Wagner type relaxations are not observed and all materials present a similar permittivity above the maximum at 300°C.

The submicron and nanostructured ceramics only presented incipient switching at 2 kV mm[-1], which was a field high enough to saturate the loop of a coarse grained material. At higher fields, the maximum and remnant polaristions increased with the field with no signs of saturation up to 4 kV mm[-1]. This indicates the ferroelectric nature of the materials, which is in agreement with a previous Rietveld study of XRD data for $0.8PbMg_{1/3}Nb_{2/3}O_3$-$0.2PbTiO_3$ at the relaxor edge of the MPB region that showed an evolution from ferroelectric monoclinic *Cm* to rhombohedral *R3m* with decreasing grain size down to 90 nm.[12] It also indicates the modification of the dynamics of domains with the decrease in grain size. A distinctive evolution of the polar domain configuration of $0.65PbMg_{1/3}Nb_{2/3}O_3$-$0.35PbTiO_3$ materials with grain size, from micron-sized lamellar domains for coarse grained ceramics to submicron/nanometer sized crosshatched domains for a ceramic with a grain size of 0.15 μm has been described by transmission electron microscopy (TEM).[10] The decrease of the maximum and remnant polarisations with the hot pressing temperature strongly suggests further shrinkage of the polar domains for grain sizes below 0.15 μm.

These effects were proposed in our previous work to be associated with the slowing down of the relaxor to ferroelectric transition that would cause the long time presence of intermediate domain configurations, intrinsically linked to the developing ferroelectric distortions.[10] The slowing down was experimentally observed with macroscopic elastic and dielectric measurements for materials with x=0.2 and found to occur quite sharply between 0.36 and 0.21 μm,[13] as shown in Figure 5.

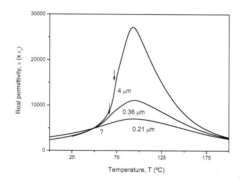

Figure 5. Temperature dependence of permittivity of $0.8PbMg_{1/3}Nb_{2/3}O_3$-$0.2PbTiO_3$ ceramics with decreasing grain size. The ferroelectric to relaxor transition is observed for the materials with 4 and 0.36 μm grain sizes, but not for the one with a grain size of 0.21 μm, in spite of all being in ferroelectric phases at room temperature.

The nanoscale domain configurations and small polarisations of the ferroelectric hysteresis loops are not promising for poling, and low piezoelectric coeefficients were indeed obtained after poling at RT. Nevertheless, we also showed in our previous work that high piezoelectric coefficients could be obtained for a materials with a grain size of 0.15 μm by applying the high electric field during cooling through the relaxor to ferroelectric transition. The same approach was used for the materials hot pressed at decreasing temperatures. Piezoelectric d_{33} coefficients were measured with a Berlincourt type meter (CP, Chesterland), and are given in Table I.

Table I. Piezoelectric coefficients and parameters of the radial resonance (frequency number and planar coupling factor) of $0.65PbMg_{1/3}Nb_{2/3}O_3$-$0.35PbTiO_3$ ceramics processed by hot pressing of the nanocrystalline powder at decreasing temperatures. Values for a coarse grained material are given for comparison.

T (°C)	d_{33} (pC N^{-1})	N_p (kHz mm)	k_p (%)	d_{31} (pC N^{-1})
900	290	2237	38	-110+3i
800	190	2339	26	-65+2i
700	80	2424	13	-35+3i
CG	525	2157	54	-160+3i

Piezoelectric radial resonances were excited for all materials, and analysed by an automatic iterative procedure that provides a set of material coefficients in complex form (ICMM, CSIC). Resonances are given in Figure 6, and the piezoelectric parameters of the submicron and nanostructured ceramics are given in Table I. There is a decrease of the piezoelectric activity with grain size, yet significant coefficients are still obtained for the materials.

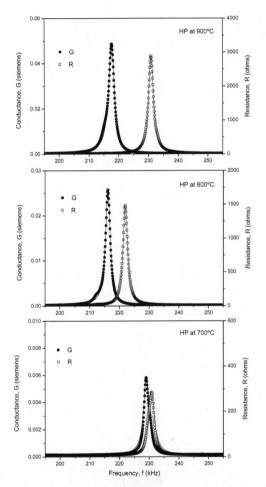

Figure 6. Piezoelectric radial resonances of $0.65PbMg_{1/3}Nb_{2/3}O_3\text{-}0.35PbTiO_3$ ceramics processed by hot pressing of the nanocrystalline powder at decreasing temperatures. Solid lines are the calculated profiles by using the coefficients obtained of the analysis, and the good agreement with the experimental resonances is a test of the reliability of the coefficients.

We previously proposed that the application of an electric field during the transition speeded up its kinetics and resulted in coarser domain configuration that resulted in significant piezoelectricity. This concept is further supported by the modification of the temperature dependence of permittivity after poling, in which the ferroelectric to relaxor transition is now observed, as can be seen in Figure 7.

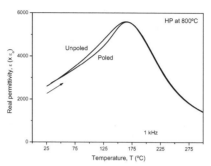

Figure 7. Temperature dependence of permittivity of a submicron- structured $0.65PbMg_{1/3}Nb_{2/3}O_3$-$0.35PbTiO_3$ ceramics before and after poling.

Lower grain sizes deeper in the nanoscale cannot be processed by this approach, and new sintering concepts need to be considered for extending the size studies down to a few tens of nm.

NANOSTRUCTURED CERAMICS OF MPB $PbZn_{1/3}Nb_{2/3}O_3$-$PbTiO_3$ AND $PbZn_{1/3}Nb_{2/3}O_3$-$PbFe_{1/2}Nb_{1/2}O_3$-$PbTiO_3$

Nanocrystalline Powder

Nanocrystalline powders of $0.92PbZn_{1/3}Nb_{2/3}O_3$-$0.08PbTiO_3$, $0.82PbZn_{1/3}Nb_{2/3}O_3$-$0.10PbFe_{1/2}Nb_{1/2}O_3$-$0.08PbTiO_3$ and $0.52PbZn_{1/3}Nb_{2/3}O_3$-$0.40PbFe_{1/2}Nb_{1/2}O_3$-$0.08PbTiO_3$ were synthesised by mechanochemical activation of analytical grade PbO, ZnO, TiO_2 and Nb_2O_5. The activation was carried out with a high energy planetary mill (Model Pulverisette 6, Fritsch) and stainless steel milling media. Details of the procedure and of the mechanisms taking place during the process can be found elsewhere.[14] The final powder consisted of only a nanocrystalline perovskite phase with ~ 12 nm crystal size after activation under tailored conditions. A non-negligible Fe (observed by EDS in TEM) contamination was found. This technique, along with high pressure synthesis[15] are the only ones that have succeeded in obtaining the $0.92PbZn_{1/3}Nb_{2/3}O_3$-$0.08PbTiO_3$ low tolerance factor perovskite.[16]

Processing by Spark Plasma Sintering

The nanocrystalline $0.92PbZn_{1/3}Nb_{2/3}O_3$-$0.08PbTiO_3$ perovskite was not stable under heating, and decomposed into a pyrochlore phase above 400°C, which prevented the processing of ceramics by conventional means. An alternative is spark plasma sintering (SPS), in which sintering is activated by a pulsed direct current and uniaxial pressure, which generally results in heating rates up to 600°C min^{-1}, and fast densification in short times (a few minutes) and low temperatures.[17] The combination of nanocrystalline powder obtained by mechanosynthesis and spark plasma sintering has been demonstrated for the processing of fine grained ceramics of a range of ferroelectric perovskites, and for the study of grain size effects.[18] In the case of $0.92PbZn_{1/3}Nb_{2/3}O_3$-$0.08PbTiO_3$, this is also advantageous for minimising the thermal decomposition of the perovskite.

SPS experiments were carried out at PNF2/CNRS-MHT, Université Paul Sabatier, Toulouse, France (Model 2080, Sumimoto). A cylindrical graphite die with an inner diameter of 8 mm was filled with 1 g of the nanocrystalline powder. A pressure of 100 MPa was applied, and high current pulses were circulated, such as a heating rate of 100°C min^{-1} resulted. Time the final sintering conditions was 3 min. An example is shown in Figure 8. More details can be found elsewhere.[19]

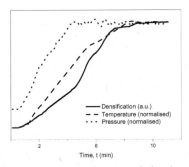

Figure 8. Sample shrinkage, temperature and pressure during the spark plasma sintering of nanocrystalline $0.92PbZn_{1/3}Nb_{2/3}O_3$-$0.08PbTiO_3$ powder at 650°C with 100 MPa.

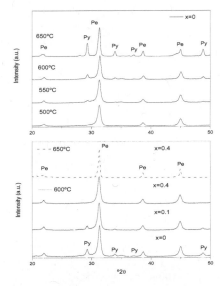

Figure 9. XRD patterns of $(0.92\text{-}x)PbZn_{1/3}Nb_{2/3}O_3$-$x\ PbFe_{1/2}Nb_{1/2}O_3$-$0.08PbTiO_3$ ceramics processed by spark plasma sintering of the nanocrystalline powder at increasing temperatures.

Ceramic materials with densification around 95% were obtained at 650°C. However, materials presented a significant amount of pyrochlore phase due to the onset of the perovskite decomposition at 550°C. This is illustrated in Figure 9, where XRD patterns of ceramics spark plasma sintered at increasing temperatures are shown. This effect prevented the temperature of the SPS to be further increased above 650°C. This is not only because the appearance of the pyrochlore phase, but also because of the die failure caused by the volume expansion at the perovskite to pyrochlore transition.[20]

An effective means of increasing the perovskite stability is by addition of $PbFe_{1/2}Nb_{1/2}O_3$ to form the $PbZn_{1/3}Nb_{2/3}O_3$-$PbFe_{1/2}Nb_{1/2}O_3$-$PbTiO_3$ ternary system, which also presents a MPB with high piezoelectric activity.[21] XRD patterns of materials of the ternary system spark plasma sintered at increasing temperatures are also shown in Figure 9, which demonstrate that single phase materials can be obtained at 650°C.

Grain size in the materials was studied by transmission electron microscopy (120 kV, CM12ST, Philips). No differences with the $PbFe_{1/2}Nb_{1/2}O_3$ content were found for a given temperature, and grain size increased from ~15, to 20 and 40 nm for the ceramics SPS at 550, 600 and 650°C, respectively. Images are shown in Figure 10.

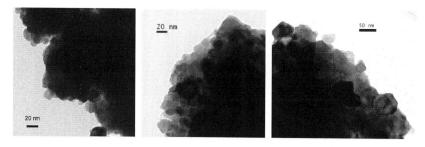

Figure 10. TEM images of $(0.92-x)PbZn_{1/3}Nb_{2/3}O_3$-$x$ $PbFe_{1/2}Nb_{1/2}O_3$-$0.08PbTiO_3$ ceramics processed by spark plasma sintering of the nanocrystalline powder. From left to right: x=0 at 550°C, x=0.1 at 600°C, x=0.4 at 650°C.

Electrical Properties

Measuring conditions were analogous to those used for the hot pressed $PbMg_{1/3}Nb_{2/3}O_3$-$PbTiO_3$ materials, but for the electrodes. Silver layers were painted and sintered at 650°C on the ceramic samples processed by hot pressing, while Pt electrodes were deposited by sputtering on the spark plasma sintered $(0.92-x)PbZn_{1/3}Nb_{2/3}O_3$-$x$ $PbFe_{1/2}Nb_{1/2}O_3$-$0.08PbTiO_3$ specimens, and annealed at only 500°C. This was necessary not only to avoid grain growth, but also to prevent perovskite decomposition in the $PbZn_{1/3}Nb_{2/3}O_3$-$PbTiO_3$ materials.

All ceramics spark plasma sintered at 550 and 600°C presented relaxor type behaviour as it is illustrated in Figure 11 for a pyrochlore free material. These materials showed densifications of ~66 and 80%.

It appeared that two different relaxor type dielectric relaxations dominated the real and imaginary components of permittivity, respectively. Both relaxations fit well to the Vogel-Fulcher relationship as shown in Figure 12, which indicates that they most probably result of the interaction among PNRs characteristic of the relaxor state. Their independent nature is suggested by the very different parameters that are obtained from the fit: a freezing temperature of 247 K, an activation energy of 34 meV and a pre-exponential factor of 2.1×10^{12} for relaxation 1, and values of 60 K, 40 meV and 3.4×10^{10} for relaxation 2. This suggest correlations for process 2 to be weaker than for process 1, and we proposed in a previous letter that they could be associated with intragranular and intergranular correlations among PNRs, respectively.[22]

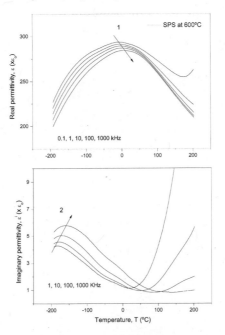

Figure 11. Temperature and frequency dependences of the real and imaginary permittivities of a $0.52PbZn_{1/3}Nb_{2/3}O_3$-$0.4PbFe_{1/2}Nb_{1/2}O_3$-$0.08PbTiO_3$ ceramic spark plasma sintered at 600°C. Two relaxations labeled as 1 and 2 are observed.

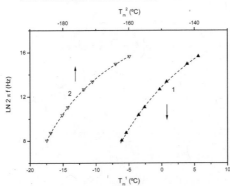

Figure 12. Fit of the two relaxation processes to a Vogel-Fulcher relationship.

The simultaneous observation of both processes in a sample is unusual, and most probably a consequence of the distinctive features of the material. On the one hand, the material showed a lognormal distribution of grain sizes with an average value of 20 nm, and that spread from 10 to 50 nm so as half of the volume was formed by grains with a size above 32 nm. Several PNRs can exist in these grains, which interaction gives place to process 1. The relaxor state is altered in relation to the high temperature relaxor state in single crystals, for the maximum in permittivity is shifted from ~170°C down to below RT. This is most probably an effect of the small number of interacting units and of the grain boundary. Also, half the materials is formed by smaller grains that can only contain one PNR. Previously reported nanoscale materials were poorly densified and no freezing was observed indicating vanishing correlations.[23] In our case, however, highly densified areas exist in the material with an average densification of 80%,[19] within which intergranular interaction might be possible, yet correlations are weakened as indicated by the lower freezing temperature.

Permittivity was also characterised for the materials SPS at 650°C, but giant permittivity with a step-like increase with temperature was found, most probably associated with Maxwell-Wagner type polarisation relaxation caused by the activation of charge carriers, as can be seen in Figure 13.

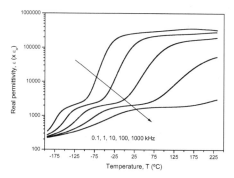

Figure 13. Permittivity of a $0.52PbZn_{1/3}Nb_{2/3}O_3$-$0.4PbFe_{1/2}Nb_{1/2}O_3$-$0.08PbTiO_3$ ceramic spark plasma sintered at 650°C.

Results on the spark plasma sintered $PbZn_{1/3}Nb_{2/3}O_3$-$PbFe_{1/2}Nb_{1/2}O_3$-$PbTiO_3$ nanostructured materials add up to the results on the hot pressed $PbMg_{1/3}Nb_{2/3}O_3$-$PbTiO_3$ submicron and nanostructured ceramics to draw a consistent picture of the grain size effects in relaxor based MPB materials from the submicron range to the nanoscale down to 15 nm.[10,13,24] Final states are controlled by the kinetics of the relaxor to ferroelectric transition that is slowed down with the decrease of grain size down to the stabilisation of the high temperature relaxor state at RT for a few tens of nm. At intermediate grain sizes a range of intermediate, submicron/nanoscale polar domain configurations exist, which can be coarsened by tailoring poling to give significant piezoelectricity for grain sizes around 100 nm. It is basically the same phenomenology that has been described with varying compositions across the solid solutions towards the relaxor edge.

NANOSTRUCTURED CERAMICS OF MPB $BiScO_3$-$PbTiO_3$

Nanocrystalline Powder

Nanocrystalline powders of $(1-x)BiSc_3O_3$-$xPbTiO_3$ with $0.6<x<0.65$ were synthesised by mechanochemical activation of analytical grade Bi_2O_3, Sc_2O_3, PbO and TiO_2. Activation was carried out with the Pulverisette 6 high energy planetary mill and stainless steel milling media. Details of the procedure and of the mechanisms taking place during the process can be found elsewhere.[21] Final powder consisted of only a nanocrystalline perovskite phase with ~ 11 nm crystal size under tailored activation conditions. Again, a non-negligible Fe contamination resulted. This technique has succeeded in obtaining nanocrystalline powders of $(1-x)BiSc_3O_3-xPbTiO_3$ with x as low as 0.2, which is far beyond phases that can be synthesised by solid state reactions. These experiments along with those in the $PbZn_{1/3}Nb_{2/3}O_3-PbTiO_3$ and $PbZn_{1/3}Nb_{2/3}O_3-PbFe_{1/2}Nb_{1/2}O_3-PbTiO_3$ systems show that mechanochemical activation is a poweful technique for the synthesis of low tolerance factor perovskites.[21]

Processing by Spark Plasma Sintering

The nanocrystalline BiScO₃-PbTiO₃ perovskite was stable under heating, so experiments at increasing temperatures were carried out until full densification was obtained at 650°C with 75 MPa. XRD confirmed perovskite single phase. An average grain size of 80 nm resulted as shown in Figure 14.

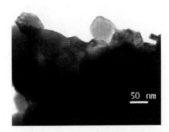

50 nm

Figure 14. TEM images of a 0.37BiScO₃-0.63PbTiO₃ ceramic
processed by spark plasma sintering of the nanocrystalline powder at 650°C.

Electrical Properties

Pt electrodes deposited by sputtering and annealed at 500°C were used. Typical temperature dependence of permittivity for the nanostructured ceramics is shown in Figure 15 at several frequencies. There was a high temperature dielectric relaxation at the temperature range in which the ferroelectric transition was expected. Nevertheless, this relaxation was already shifted above 500°C for frequencies higher than 10 kHz, and the transition anomaly could be observed at the highest frequencies at approximately the same temperature that in coarse grained materials; 442°C, yet broadened and flattened.[25]

The ferroelectric nature of the material was corroborated by ferroelectric hysteresis loop measurements. A loop after compensation is shown in Figure 16 that presented a remnant polarisation of 19 μC cm⁻², which has to be compared with 32 μC cm⁻² for a coarse grained material.[25]

Furthermore, the nanostructured ceramics could be poled, and a d_{33} of ~40 pC N⁻¹ was obtained after poling at 150°C with 4 kV mm⁻¹. Values above 150 pC N⁻¹ and up to 330 pC N⁻¹ are obtained for coarse grained ceramics of BiScO₃-PbTiO₃ in the MPB region. Piezoelectric radial resonances were excited and analysed, an example of which is shown in Figure 17. A frequency number of 2488 kHz mm, a planar coupling factor of 9%, and a d_{31} coefficient of -14 pC N⁻¹ were obtained.[25]

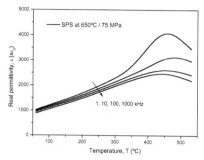

Figure 15. Temperature dependence of permittivity of a $0.375BiScO_3$-$0.625PbTiO_3$ ceramic processed by spark plasma sintering of the nanocrystalline powder at 650°C.

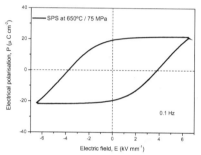

Figure 16. Ferroelectric hysteresis loop of a $0.375BiScO_3$-$0.625PbTiO_3$ ceramic processed by spark plasma sintering of the nanocrystalline powder at 650°C.

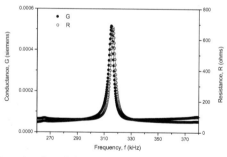

Figure 17. Piezoelectric radial resonance a $0.37BiScO_3$-$0.63PbTiO_3$ ceramic processed by spark plasma sintering of the nanocrystalline powder at 650°C.

Efforts are currently focused on obtaining materials with grain size as small as possible, while maintaining high densifications. Ceramics with an average grain size of ~20 nm have been obtained by tailoring the temperature and pressure of the SPS. Preliminar electrical measurements indicate they might be non-ferroelectric.

CONCLUSIONS

Activities at ICMM on the processing of submicron and nanostructured ceramics of high sensitivity piezoelectric materials, and on the study of their properties have been reviewed and updated. Nanocrystalline powders obtained by mechanosynthesis have been shown to be very suitable for the processing of these materials, even of low tolerance factor perovskites. Hot pressing of a nanocrystalline powder so synthesised has allowed the processing of relaxor based MPB $PbMg_{1/3}Nb_{2/3}O_3$-$PbTiO_3$ ceramics with grain size in the submicron range down to the nanoscale (90 nm). Deeper in the nanoscale, spark plasma sintering of the same type of nanocrystalline powder succeeded in processing materials of $PbZn_{1/3}Nb_{2/3}O_3$-$(PbFe_{1/2}Nb_{1/2}O_3)$-$PbTiO_3$ and $BiScO_3$-$PbTiO_3$ with a grain size as small as 15 nm.

The study of the electrical properties of the different relaxor based MPB materials has allowed describing the grain size effects in this type of materials down to a few tens of nm. Size effects seem to be associated with the slowing down of the kinetics of the relaxor to ferroelectric transition in the submicron range. This causes a range of intermediate submicron/nanoscale polar domain configurations to stabilize at RT. At grain sizes of a few tens of nm, kinetics is such that the high temperature relaxor state persists down to 77 K. Its dynamics has been investigated, and evidence of correlations among PNRs across grain boundaries has been found.

Finally, functionality has demonstrated for MPB $PbMg_{1/3}Nb_{2/3}O_3$-$PbTiO_3$ with grain size of 90 nm after tailored poling, and for MPB $BiScO_3$-$PbTiO_3$ with 80 nm.

ACKNOWLEDGMENTS

Research on $PbMg_{1/3}Nb_{2/3}O_3$-$PbTiO_3$ is a collaboration between ICMM and IJS, where the nanocrystalline powder was developed. It is part of the Joint Programme of Research of the European Network of Excellence MIND (Ref. NoE 515757-2). National funding by MICINN is acknowledged (Ref. MAT2008-02003/NAN).

Research on the synthesis of low tolerance factor perovskites has been entirely carried out at ICMM with national projects (Refs. MAT2007-61884 and MAT2008-02003/NAN). Spark plasma sintering experiments were done in collaboration with CEMES (CNRS) in the framework of the ESF COST Action 539 ELENA. Technical support by Ms. I. Martínez is also acknowledged.

REFERENCES

[1]C. Pithan, D. Hennings and R. Waser, Progress in the synthesis of nanocrystalline $BaTiO_3$ powders for MLCC, *Int. J. Appl. Ceram. Technol.* **2**, 1-14 (2005).
[2]P. Muralt, Recent progress in materials issues for piezoelectric MEMS, *J. Am. Ceram. Soc.* **91**, 1385-1396 (2008).
[3]M.T. Buscaglia, M. Viviani, V. Buscaglia, L. Mitoseriu, A. Testino, P. Nanni, Z. Zhao, M. Nygren, C. Harneaga, D. Piazza and C. Galassi, High dielectric constant and frozen macroscopic polarisation in dense nanocrystalline $BaTiO_3$ ceramics, *Phys. Rev. B* **73**, art. n° 064114 (2006).
[4]C.A. Randall, N. Kim, J.P. Kucera, W. Cao and T.R. Shrout, Intrinsic and extrinsic size effects in fine grained morphotropic phase boundary lead zirconate ceramics, *J. Am. Ceram. Soc.* **81**, 677-688 (1998).
[5]S.E. Park and T.R. Shrout, Ultrahigh strain and piezoelectric behaviour in relaxor based ferroelectric single crystals, *J. Appl. Phys.* **82**, 1804-1811 (1997).
[6]R.E. Eitel, C.A. Randall, T.R. Shrout and S.E. Park, Preparation and characterisation of high temperature perovskite ferroelectrics in the solid solution (1-x)$BiScO_3$-x$PbTiO_3$, *Jpn. J. Appl. Phys.* **41**, 2099-2104 (2002).
[7]D. Kuscer, J. Holc and M. Kosec, Formation of $0.65PbMg_{1/3}Nb_{2/3}O_3$-$0.35PbTiO_3$ using a high energy milling process, *J. Am. Ceram. Soc.* **90**, 29-35 (2007).

[8]M. Algueró, A. Moure, L. Pardo, J. Holc and M. Kosec, Processing by mechanosynthesis and properties of piezoelectric $PbMg_{1/3}Nb_{2/3}O_3$-$PbTiO_3$ with different compositions, *Acta Mater.* **54**, 501-511 (2006).

[9]H. Amorín, J. Ricote, R. Jiménez, J. Holc, M. Kosec and M. Algueró, Submicron and nanostructured $0.8PbMg_{1/3}Nb_{2/3}O_3$-$0.2PbTiO_3$ ceramics by hot pressing of nanocrystalline powders, *Scripta Mater.* **58**, 755-758 (2008).

[10]M. Algueró, J. Ricote, R. Jiménez, P. Ramos, J. Carreaud, B. Dkhil, J.M. Kiat, J. Holc and M. Kosec, Size effect in morphotropic phase boundary $PbMg_{1/3}Nb_{2/3}O_3$-$PbTiO_3$, *Appl. Phys. Lett.* **91**, art. nº 112905 (2007).

[11]V. Botvun, S. Kamba, S. Veljko, D. Nuzhnyy, J. Kroupa, M. Savinov, P. Vanek, J. Petzelt, J. Holc, M. Kosec, H. Amorín and M. Algueró, Broadband dielectric spectroscopy of phonons and polar nanoclusters in $PbMg_{1/3}Nb_{2/3}O_3$-$PbTiO_3$ ceramics: grain size effects, *Phys. Rev. B* **79**, art. nº 104111 (2009).

[12]J. Carreaud, J.M. Kiat, B. Dkhil, M. Algueró, J. Ricote, R. Jiménez, J. Holc and M. Kosec, Monoclinic morphotropic phase and grain size induced polarization rotation in $PbMg_{1/3}Nb_{2/3}O_3$-$PbTiO_3$, *Appl. Phys. Lett.* **89**, art. nº 252906 (2006).

[13]R. Jiménez, H. Amorín, J. Ricote, J. Carreaud, J.M. Kiat, B. Dkhil, J. Holc, M. Kosec and M. Algueró, Effect of grain size on the transition between ferroelectric and relaxor states in $0.8PbMg_{1/3}Nb_{2/3}O_3$-$0.2PbTiO_3$ ceramics, *Phys. Rev. B* **78**, art. nº 094103 (2008).

[14]M. Algueró, J. Ricote and A. Castro, Mechanosynthesis and thermal stability of piezoelectric perovskite $0.92PbZn_{1/3}Nb_{2/3}O_3$-$0.08PbTiO_3$ powders, *J. Am. Ceram. Soc.* **87**, 772-778 (2004).

[15]Y. Matsuo, H. Sasaki, S. Hayakawa, F. Kanamaru and M. Koizumi, High pressure synthesis of perovskite type $Pb(Zn_{1/3}Nb_{2/3})O_3$, *J. Am. Ceram. Soc.* **52**, 516-517 (1969).

[16]A. Halliyal, U. Kumar, R.E. Newham and L.E. Cross, Stabilization of the perovskite phase and dielectric properties of ceramics in the $Pb(Zn_{1/3}Nb_{2/3})O_3$-$BaTiO_3$ system, *Am. Ceram. Soc. Bull.* **66**, 671-676 (1987).

[17]Z.A. Munir, U. Anselmi-Tamburini and M. Ohyanagi, Title: The effect of electric field and pressure on the synthesis and consolidation of materials: A review of the spark plasma sintering method, *J. Mater. Sci.* **41**, 763-777 (2006).

[18]T. Hungría, J. Galy and A. Castro, The Spark Plasma Sintering as a useful technique to the nanostructuration of piezo-ferroelectric materials, *Adv. Eng. Mater.* **11** 615-631 (2009).

[19]T. Hungría, H. Amorín, J. Galy, J. Ricote, M. Algueró and A. Castro, Nanostructured ceramics of $0.92PbZn_{1/3}Nb_{2/3}O_3$-$0.08PbTiO_3$ processed by SPS of nanocrystalline powders obtained by mechanosynthesis, *Nanotechnol.* **19**, art. nº 155609 (2008).

[20]T. Hungría, A. Castro, M. Algueró and J. Galy, Uncontrollable expansion of $Pb(Zn_{1/3}Nb_{2/3})O_3$-$PbTiO_3$ perovskite $\Rightarrow$ pyrochlore transition during spark plasma sintering; mechanism proposal using infinite periodic minimal surfaces, *J. Solid State Chem.* **181**, 2918-2923 (2008).

[21]M. Algueró, J. Ricote, T. Hungría and A. Castro, High sensitivity piezoelectric, low-tolerance-factor perovskites by mechanosynthesis, *Chem. Mater.* **19**, 4982-4990 (2007).

[22]H. Amorín, R. Jiménez, T. Hungría, A. Castro and M. Algueró, Relaxor behaviour in nanostructured $Pb(Zn_{1/3}Nb_{2/3})O_3$-$Pb(Fe_{1/2}Nb_{1/2})O_3$-$PbTiO_3$ ceramics, *Appl. Phys. Lett.* **94**, art. nº 152902 (2009).

[23]J. Carreaud, P. Gemeiner, J.M. Kiat, B. Dkhil, C. Bogicevic, T. Rojac and B. Malic, Size driven relaxation and polar states in $PbMg_{1/3}Nb_{2/3}O_3$ based systems, *Phys. Rev. B* **72**, art. nº 174115 (2005).

[24]M. Algueró, T. Hungría, H. Amorín, J. Ricote, J. Galy and A. Castro, Relaxor behaviour, polarisation build up, and switching in nanostructured $0.92PbZn_{1/3}Nb_{2/3}O_3$-$0.08PbTiO_3$ ceramics, *Small* **3**, 1906-1911 (2007).

[25]M. Algueró, H. Amorín, T. Hungría, J. Galy and A. Castro, Macroscopic ferroelectricity and piezoelectricity in nanostructured $BiScO_3$-$PbTiO_3$ ceramics, *Appl. Phys. Lett.* **94**, art. nº 012902 (2009).

TRANSFORMATION OF CURRENT LIMITING EFFECT INTO VARISTOR EFFECT IN TIN DIOXIDE BASED CERAMICS

A. N. Bondarchuk, A. B. Glot, M. Marquez Miranda

Universidad Tecnologica de la Mixteca, Huajuapan de Leon,
Oaxaca 69000 Mexico, e-mail: alexbond@mixteco.utm.mx

ABSTRACT
The current limiting effect (current is increased weaker than voltage, saturated and even decreased) and its transformation into the varistor effect were found in SnO_2–Co_3O_4–Nb_2O_5–Cr_2O_3 ceramics sintered at relatively low temperatures 1100–1200°C. Electrical measurements and scanning electron microscopy suggest that this effect is related to electrical conduction controlled by grain-boundary potential barriers.

INTRODUCTION
At the present time zinc oxide (ZnO) based varistor ceramics are widely used for absorption of transient surges in electrical and electronic equipment.[1] This application is based on the high nonlinearity (β >40) of the superlinear and symmetric current-voltage characteristics observed in zinc oxide based ceramics.[1-4] Such varistor effect was observed in other oxide materials based on TiO_2,[5] WO_3,[6] and SnO_2.[7] Though, the nonlinearity coefficient $\beta = (U/I)(dI/dU)$ in such materials was low ($\beta \leq 10$). Later quite high nonlinearity was found in tin dioxide based varistor ceramics in a system SnO_2–Bi_2O_3-Co_3O_4-BaO-Nb_2O_5 ($\beta = 21$)[8] and in a system SnO_2–Co_3O_4–Nb_2O_5–Cr_2O_3 (β $\cong 40$).[9] The varistor effect in oxide ceramics is related to the decrease in the barrier height with increasing electric field.[10-13] Some results in the area of oxide varistor ceramics were outlined in the reviews.[3,4,13,14]

Therefore, SnO_2–Co_3O_4–Nb_2O_5–Cr_2O_3 ceramics are considered traditionally as a model SnO_2 varistor material with relatively high nonlinearity coefficient (β $\cong 40$). However, it was recently observed that SnO_2-Co_3O_4-Nb_2O_5-Cr_2O_3 ceramics sintered at relatively low temperatures 1100-1200°C exhibit strongly different non-Ohmic behavior: current is increased weaker than voltage, saturated and even decreased.[15] Such current limiting effect looks quite interesting.

In this paper the electrical properties of SnO_2–Co_3O_4–Nb_2O_5–Cr_2O_3 ceramics sintered at 1100-1200°C are studied and the nature of observed current limiting effect is discussed.

EXPERIMENTAL PROCEDURE
Ceramic samples (mol. %) 98.9SnO_2-1Co_3O_4-0.05Nb_2O_5-0.05Cr_2O_3 were obtained by mixed oxide route[15] with sintering at 1100-1400°C (100 min) in air with heating rate of 5°C /min and cooling rate of 2°C /min. The thickness of sintered samples was 0.07-0.1 cm. Ag electrodes were obtained at 800°C (10 min, 2°C /min).

Current-voltage characteristics were recorded in air using source-measure unit Keithley 237 controlled by computer. Rectangular unipolar voltage pulses with duration of 100 ms, 100 ms pause between pulses were used. Current was measured at the end of each voltage pulse.

Capacitance C and ac conductance G were measured in parallel circuit using computer controlled QuadTech 7600 LCR meter with ac voltage amplitude 0.5 V. The dc term was not subtracted from the total dielectric losses. Small-signal capacitance $C(U)$ and ac conductance $G(U)$ versus dc voltage (Keithley 6487) were obtained at 100 Hz using rectangular unipolar voltage pulses with duration of 4s, 4s interval between pulses, and 0.1V amplitude increment.

To obtain $C(U)$ and $G(U)$ dependences for ceramics, where current limiting effect is observed, fresh samples (not subjected to the influence of electric field) were used. For such experiment ceramic tablet was broken for several parts. Then $I(U)$ dependence for one such a part from each tablet was recorded and current limiting behavior was confirmed.

The relative dielectric permittivity ε and the ac conductivity g were calculated from the formulas $C = \varepsilon \varepsilon_0 S / d$ and $G^{-1} = g^{-1} d / S$, where ε_0 is the permittivity of vacuum, d and S are the thickness and the cross-section of a sample, respectively.

RESULTS

Current-voltage dependence

The dependences of current density on electric field ($j(E)$) in SnO_2–Co_3O_4–Nb_2O_5–Cr_2O_3 ceramics sintered at relatively low temperature 1100°C are shown in Fig.1. Initial $j(E)$ curve contains linear, superlinear ($\beta > 1$) and sublinear ($\beta < 1$) regions at increase in voltage (Fig.1a, curve 1). At the sublinear region current is increased weaker than voltage, saturated and even decreased with voltage. Such a behavior let us consider as the current limiting effect. However, at voltage decrease only varistor behavior is observed and $j(E)$ curve is displaced to lower current values (Fig.1a, curve 2). It was observed for some next cycles of rise and drop of voltage.

After the numerous (5-10) cycles of rise and drop of voltage at the same polarity the current limiting effect is practically disappeared and only linear and superlinear regions are seen (Fig.1a, curves 3 and 4). Some difference between the curves obtained at increase and decrease in voltage takes place (Fig.1a, curves 3 and 4). But both curves show varistor behaviour. Therefore, the current limiting effect in tin dioxide ceramics is transformed into the varistor effect.

The observed current limiting effect can be restored by the heat treatment at elevated temperatures (about 800°C) or by the reversing of voltage polarity. The $j(E)$ dependences recorded at the opposite polarity in the same sample are presented in Fig.1a (curves 5 and 6). The character of $j(E)$ dependences obtained at both polarities is similar (Fig.1a, curves 1-2 and 5-6).

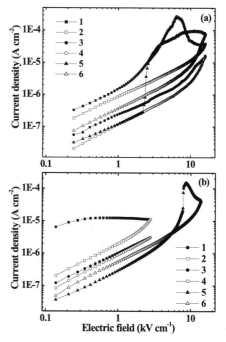

Figure 1. Current density versus electric field for two samples of SnO_2-Co_3O_4-Nb_2O_5-Cr_2O_3 ceramics sintered at 1100°C. Data were recorded at increase (1, 3, 5) and decrease (2, 4, 6) in voltage. Two groups of curves (1-4 and 5-6) were obtained at opposite voltage polarity.

The shape of $j(E)$ curve with sublinear region can be varied. The sublinear region can be observed after the superlinear (Fig.1a, curve 1) or after the linear region (Fig.1b, curve 1). Besides that, after the numerous cycles of rise and drop of voltage the sublinear region can be shifted to higher voltages (Fig.1b, curve 5).

Time dependences of current density are presented in Fig.2. At low electric field the current density is decreased on time and is not depended on voltage polarity (Fig.2, curves 1 and 2). However, at higher electric fields the reverse of voltage polarity causes a rise and subsequent drop in the current density (Fig.2, curves 4-6). The additional experiment shows that without the reverse of voltage polarity only the decrease in current density is observed.

In spite of the obvious degradation the observed current limiting effect looks interesting and unexpected because in $SnO_2-Co_3O_4-Nb_2O_5-Cr_2O_3$ ceramics usually varistor effect is observed.[9,14] The reason of such a difference in non-Ohmic behavior can be seen in the low sintering temperature used here. Our experiments show that $SnO_2-Co_3O_4-Nb_2O_5-Cr_2O_3$ ceramics of the same composition but sintered at 1300-1400°C exhibit only varistor behavior according to the published data.[9,14]

The variation in the sintering temperature gives alterations in the microstructure. Ceramic samples obtained at 1100-1200°C (with current limiting effect) are not sintered completely and many pores are seen (Fig.3a and Fig.3b). However, ceramics sintered at higher temperatures exhibit the higher density and the larger grain size (Fig.3c).

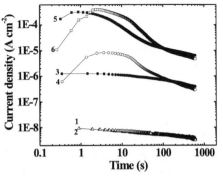

Figure 2. Dependences of current density on time in $SnO_2-Co_3O_4-Nb_2O_5-Cr_2O_3$ ceramics sintered at 1100°C. The data were recorded at electric field 0.01 kV/cm (curves 1 and 2) then at 1 kV/cm (curves 3 and 4) and finally at 10 kV/cm (curves 5 and 6). Curves 2, 4, 6 were recorded at opposite voltage polarity in respect of curves 1, 3, 5.

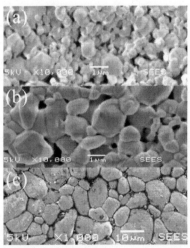

Figure 3. SEM micrographs of $SnO_2-Co_3O_4-Nb_2O_5-Cr_2O_3$ ceramics sintered at 1100°C (a), 1200°C (b) and 1400°C (c).

Dielectric properties

It would be interesting to explore the behavior of capacitance and conductance in ceramics where under the influence of the electric field the current limiting effect is transformed into the varistor effect.

The dependences of the relative dielectric permittivity ε and ac conductivity g at different frequencies on electric field are shown in Fig.4 and Fig.5. Data for three consecutive measurements at 100 Hz are presented in Fig.4 (curves 1-3). The relative dielectric permittivity is increased and further decreased with electric field (Fig.4a, curve 1). The ac conductivity is decreased with electric field (Fig.4b, curve 1). While consequent measuring at fixed frequency, the relative dielectric permittivity ε and ac conductivity g are irreversibly shifted to lower values (curves 1-3 in Fig.4).

The dependences of ε and g on dc voltage obtained at 100, 400 and 1000 Hz in the same sample are shown in Fig.5. With increasing frequency the decrease in ε and g is observed (Fig.5). Here as well the irreversible changes take place. As a result, the curves 4 in Fig.5 obtained at 100 Hz

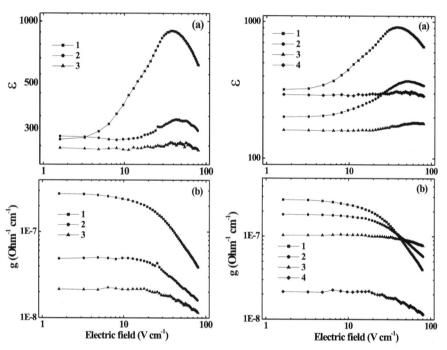

Figure 4. (a) The relative dielectric permittivity and (b) a.c. conductivity versus electric field for SnO_2-Co_3O_4-Nb_2O_5-Cr_2O_3 ceramics sintered at 1100°C. Curves 1-3 were obtained at consecutive measurements.

Figure 5. (a) The relative dielectric permittivity and (b) a.c. conductivity versus electric field for SnO_2-Co_3O_4-Nb_2O_5-Cr_2O_3 ceramics sintered at 1100°C. Data were obtained during the consecutive measurements at 100Hz (curve 1), 400Hz (curve 2), 1000Hz (curve 3), and at 100Hz (curve 4).

are shifted in respect of the initial curve 1 in Fig.5 obtained at 100 Hz.

It should be noted that before and after ac measurements (Fig.4 and Fig.5) current–voltage dependences in the same sample were recorded. They are shown in Fig.6. Therefore, the ac data in Fig. 4 and Fig.5 are obtained for ceramics with current limiting effect.

After the transformation of the current limiting effect into the varistor effect the relative dielectric permittivity and the ac conductivity become quite low ($\varepsilon \cong 15-20$, $g \cong 1\ 10^{-9} Ohm^{-1} \cdot cm^{-1}$) and their dependence on electric field is weak.

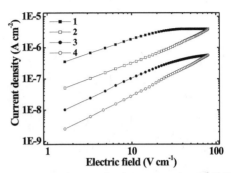

Figure 6. Current density versus electric field recorded before (curves 1-2) and after (curves 3-4) ac characteristics presented in Fig.4 and Fig.5. Data were obtained at increase (1, 3) and decrease (2, 4) in voltage.

DISCUSSION

Obtained results and the known fundamental properties of oxide semiconductor ceramics suggest that electrical conduction mechanism in studied ceramics can be understood in the frames of the grain boundary controlled conduction. Observed current limiting behavior in SnO₂-based ceramics can be explained by the additional oxygen adsorption in electric field and capture of electrons at the grain-boundary (GB) states.[15,16]

The current limiting effect is transformed into the varistor effect because after the numerous cycles of rise and drop of voltage the GB states are filled. The barrier height becomes higher and $j(E)$ dependence is shifted to lower current densities (Fig. 1, curves 3 and 4). The additionally captured electrons escape from the grain-boundary states at 800°C heat treatment and the next application of voltage again causes the current limiting effect.

The sensitivity of $j(E)$ behavior to the voltage polarity (Fig.1 and Fig.2) can indicate that there are two groups of GB states which act independently under applied voltage. It can be if two back-to-back connected Schottky barriers at the grain boundary are separated by relatively thick intergranular layer. The current is controlled by the forward biased barrier.

It is assumed that due to the decrease in the barrier height under applied voltage, the states related to the forward biased barrier are filled. At the same time the states related to the reverse biased barrier are emptied as a result of the decrease in the capture rate of electrons from the positive grain and tunneling of captured electrons to the conduction band of positive grain.

The change in voltage polarity leads to the filling of the states previously emptied at the preceding polarity. At the same time the states filled at the preceding polarity are emptied. This creates conditions for an appearance of the current limiting effect again (Fig. 1, curves 5).

In SnO₂–Co₃O₄–Nb₂O₅–Cr₂O₃ ceramics with varistor effect (sintered at 1300-1400°C) two groups of GB states probably are situated quite close because an intergranular layer can be very thin. Therefore, these two groups of GB states can be considered as a whole system. As a result, negative localized charge at grain boundary is changed weakly. Such a model is frequently used for varistor ceramics.[3,4,10-14,17]

Observed growth and subsequent decrease of the relative dielectric permittivity ε with electric field (Fig.4a, curve 1) at low frequency can be explained by the charge trapping at the grain boundary which modulates the thermionic emission current across the boundary.[18] In this case the decrease in ε are related to the filling of a majority of the grain-boundary traps.

The low-frequency ac conductivity is decreased at higher fields (Fig.4b, curve 1). However, theory [18] predicts that quite weak decrease in the low-frequency ac conductance at low fields is followed by its strong increase at higher fields. This difference suggests that in $SnO_2-Co_3O_4-Nb_2O_5-Cr_2O_3$ ceramics sintered at 1100°C the charge trapping at the grain boundaries is accompanied by a substantial growth in the barrier height and the width of the depletion region. Such increase in the barrier height is related to the oxygen adsorption in electric field[15,16] and leads to the current limiting effect (Fig.1, curves 1). The model in the paper[18] is more applicable for the grain boundary with varistor effect and does not consider the current limiting effect.

In general case, capacitance contains high-frequency component and frequency-decreasing term.[18] Therefore, the relative dielectric permittivity ε should decrease with frequency as in experiment (Fig.5). The high-frequency capacitance in the case of $SnO_2-Co_3O_4-Nb_2O_5-Cr_2O_3$ ceramics sintered at 1100°C is assumed to be additionally decreasing on voltage (due to the widening of the depletion region at oxygen adsorption in electric field [15,16]). Therefore, in our case with increasing electric field not only low-frequency capacitive term becomes lower (according to the work[18]) but additionally high-frequency capacitance is decreased. This detail explains why under the electric field ε and g are irreversibly shifted to lower values while consequent measurements (Fig.4 and Fig.5).

The irreversible decrease of g in electric field means the decrease of the dielectric losses (at dc and ac): $\varepsilon^{''} = g(\varepsilon_0\omega)^{-1}$, where $\varepsilon^{''}$ is the coefficient of dielectric losses, ω is the angular frequency. Such decrease of $\varepsilon^{''}$ in this material is due to the growth of the barrier height. At fixed frequency the behavior of $\varepsilon^{''}$ is analogous to the behavior of g (Fig.4). For the data in Fig.5 frequency is varied within one decade and, therefore, the behavior of $\varepsilon^{''}$ and g on electric field is quite similar.

CONCLUSION

The current limiting effect and its transformation into the varistor effect are observed in $SnO_2-Co_3O_4-Nb_2O_5-Cr_2O_3$ ceramics obtained at relatively low temperatures 1100-1200°C. This phenomenon is explained in terms of the modified grain-boundary model considering a relation between the electronic and adsorption processes. High porosity of such materials facilitates a penetration of oxygen into the bulk of ceramics.

Observed growth and subsequent decrease in the low-frequency relative dielectric permittivity with electric field is explained by the charge trapping at the grain boundary which modulates the thermionic emission current across the boundary. The rise in the barrier height (due to the charge trapping related to the oxygen adsorption in electric field) is responsible for the observed decrease in the relative dielectric permittivity and ac conductivity at low frequency.

REFERENCES

[1]L. Levinson and H. Philipp, ZnO Varistors for Transient Protection, *IEEE Trans. Parts, Hybrids, and Packaging*, **13**, 338-343 (1977).

[2]M. Matsuoka, Non-Ohmic Properties of Zinc Oxide Ceramics, *Jap. J. Appl. Phys.*, **10**, 736-746 (1971).

[3]T. K. Gupta, Application of Zinc Oxide Varistors, *J. Am. Ceram. Soc.*, **73**, 1817-1840 (1990).

[4]D. R. Clarke, Varistor ceramics, *J. Am. Ceram. Soc.*, **82**, 485-502 (1999).

[5]M. F. Yan and W. W. Rhodes, Preparation and properties of TiO_2 varistors, *Appl. Phys. Lett.*, **40**, 536-537 (1982).

[6] V. O. Makarov and M. Trontelj, Novel varistor material based on tungsten oxide, *J. Mat. Sci. Lett.*, **13**, 937-939 (1994).

[7]A. B. Glot, A. M. Chakk, B. K. Chernyi, and A. Y. Yakunin, Dependence of the electrical conductivities of the semiconductors ZnO-SnO$_2$-Bi$_2$O$_3$ on the temperature and additional heat-treatment procedure, *Inorganic Materials,* **10**, 1866-1868 (1974).

[8]A. B. Glot and A. P. Zlobin, Non-Ohmic conductivity of tin dioxide ceramics, *Inorganic Materials*, **25**, 274-276 (1989).

[9]S. A. Pianaro, P. R. Bueno, E. Longo, and J. A. Varela, A new SnO$_2$-based varistor system, *J. Mater. Sci. Lett.*, **14**, 692-694 (1995).

[10]G. D. Mahan, L. M. Levinson, and H. R. Philipp, Theory of conduction in ZnO varistors, *J. Appl. Phys.*, **50**, 2799-2812 (1979).

[11]L. K. J. Vanadamme and J. C. Brugman, Conduction mechanisms in ZnO varistors, *J. Appl. Phys.,* **51**, 4240-4244 (1980).

[12]G. E. Pike, Electronic properties of ZnO varistors: a new model, In: *Grain Boundaries in Semiconductors. Proc. Mater. Res. Soc. Ann. Meet.,* ed. G. E. Pike, C. H. Seager, H. J. Leamy, Elsevier, 369-379 (1982).

[13]A. B. Glot, Non-ohmic Conduction in Oxide Ceramics: Tin Dioxide and Zinc Oxide Varistors. In: *Ceramic Materials Research Trends*, ed. P. B. Lin, Nova Science Publishers, Inc., 227-273 (2007).

[14]P. R. Bueno, J. A. Varela, and E. Longo, SnO$_2$, ZnO and related polycrystalline compound semiconductors: An overview and review on the voltage-dependent resistance (non-ohmic) feature, *J. Eur. Ceram. Soc.*, **28**, 505-529 (2008).

[15]A. N. Bondarchuk and A. B. Glot, Transformation of current limiting effect into varistor effect in tin dioxide based ceramics, *J. Phys. D: Appl. Phys,* 2008, **41**, 175306 (3pp) doi: 10.1088/0022-3727/41/17/175306.

[16]A. Bondarchuk, A. Glot, G. Behr, and J. Werner, Current saturation in indium oxide based ceramics, *Eur. Phys. J: Appl. Phys,* **39**, 211-217 (2007).

[17]G. E. Pike and C. H. Seager, The dc voltage dependence of semiconductor grain-boundary resistance, *J. Appl. Phys.*, **50**, 3414-3422 (1979).

[18]G. E. Pike, Semiconductor grain-boundary admittance: Theory, *Phys. Rev. B*, **30**, 795-802 (1984).

FABRICATION OF MoSi$_2$–Si-COMPOSITE THIN FILMS FOR OXIDATION-RESISTANT THIN-FILM HEATERS

Teppei Hayashi, Masashi Sato, Yuuki Sato, and Shinzo Yoshikado
Department of Electronics, Doshisha University
Kyotanabe 610-0321, Japan

ABSTRACT
Thin films of silicon (Si)-added-molybdenum silicate (MoSi$_2$) were deposited on silicon nitride (Si$_3$N$_4$) and alumina substrates by radio-frequency magnetron sputtering using a target made from MoSi$_2$ and Si powders. X-ray diffraction (XRD) revealed that Si-added-MoSi$_2$ thin films consisted of a mixture of Si with a cubic structure and MoSi$_2$ with a hexagonal structure. Furthermore, scanning electron microscope (SEM) images and XRD patterns of a thin film revealed that the Si-added-MoSi$_2$ thin film deposited on an alumina substrate has a columnar structure. The Si and MoSi$_2$ particle distributions in a thin film can be determined from element mapping by an energy-dispersive X-ray spectrometer (EDS) and using a backscattered electron (BSE) detector in an SEM. However, microstructures formed by Si and MoSi$_2$ particles could not be observed at the resolution of the BSE measurements. The resistance of MoSi$_2$ thin-film heaters fabricated using a Si$_3$N$_4$ substrate heated for a long time in air increased with increasing heating time at temperatures around 420 °C due to the generation of MoO$_3$, which is an insulator, in the thin film due to the oxidation of Mo in MoSi$_2$. On the other hand, the resistance of Si-added-MoSi$_2$ thin-film heaters fabricated using a Si$_3$N$_4$ substrate was stable in air at temperatures near 480 °C over long heating times exceeding 950 h. Thus, the oxidation resistance of thin-film heaters fabricated by adding Si to MoSi$_2$ was drastically improved compared with that of MoSi$_2$ thin-film heaters. Based on these results, it is speculated that thin films of Si-added-MoSi$_2$ formed microstructures of Si particles or that the Si layer on a MoSi$_2$ particle prevented oxidation of MoSi$_2$.

INTRODUCTION

Molybdenum silicate (MoSi$_2$) has a high melting point of 2030 °C and it is widely used in heaters that can be used at high temperatures in oxidizing atmospheres, similar to silicon carbide (SiC) heaters. Molybdenum silicate heaters can be used up to 1800 °C, which is higher than the maximum operating temperature of SiC heaters. Furthermore, MoSi$_2$ has a higher chemical stability than SiC in various atmospheres. However, MoSi$_2$ is brittle at room temperature and is easily softened above 1300 °C. Therefore, new strategies are required for its installation and use in heaters[1–2]. We have evaluated heaters fabricated by depositing MoSi$_2$ on a substrate by radio-frequency (RF) magnetron sputtering[3]. Currently, the heaters are fabricated by depositing MoSi$_2$ thin films on a silicon nitride (Si$_3$N$_4$) or alumina substrate and installing silver or platinum (Pt) electrodes.

The performance of such thin-film heaters deteriorates with increasing heating time in air at high temperatures. For example, the resistance of MoSi$_2$ thin-film heaters increases with increasing heating time in air at temperatures near 300 °C, because of oxidation of Mo in MoSi$_2$[2]. On the other hand, molybdenum silicates with higher composition ratios of Si to Mo (e.g., Mo$_3$Si, Mo$_5$Si$_3$, and MoSi$_2$) are more oxidation resistant. There is no molybdenum silicate that has a higher composition ratio of Si to Mo than two for MoSi$_2$. Based on this, fabrication of silicon (Si)-added-MoSi$_2$ thin-film heaters with much higher Si contents was investigated using a sputtering target made from a mixture of Si and MoSi$_2$ powders to improve the oxidation resistance of MoSi$_2$.

The microstructures of the thin films were evaluated by composition images obtained using an energy-dispersive X-ray spectrometer (EDS) installed in a scanning electron microscope (SEM). However, the element maps obtained had low resolutions making the element distributions unclear.

Furthermore, it was difficult to observe Si-added-$MoSi_2$ thin films using a transmission electron microscope (TEM) because it requires complicated sample processing. On the other hand, a high-contrast composition image can be easily obtained using a back-scattered electron (BSE) detector with an SEM, although it is not possible to perform quantitative analysis using a BSE detector. A high-contrast composition image can be formed because the emission efficiency of back-scattered electrons increases with increasing atomic number so that each element has a different back-scattered electron intensity. Moreover, a BSE detector has a higher resolution than an EDS because the area over which fluorescent X-rays are generated by electron irradiation is larger than that of back-scattered electrons.

EXPERIMENT

Mixtures of Si and $MoSi_2$ powders were prepared by mixing Mo (purity: 99.5%) and $MoSi_2$ (purity: 99.5%) powders in an agate bowl to produce mixtures having Mo to Si molar composition ratios Mo:Si = 1:2.0, 2.2, 2.3, 2.6, 2.8, 3.0, and 3.2. The powders were subjected to a pressure of 25 MPa on the target holder of an anoxic copper plate (diameter: 100 mm, depth: 4 mm). The target holder was placed in an RF magnetron sputtering system (Anelva, SPF-210B). A substrate was placed opposite the target holder. The distance of the target from the substrate was approximately 50 mm. A Si_3N_4 substrate (50×50 mm², thickness: 2.5 mm) was used. However, an alumina substrate also was used for obtaining cross-sectional images of thin films, because the Si_3N_4 substrate was too hard to break. Hereafter, thin-film heaters fabricated using a target made from $MoSi_2$ powder is referred to as the MoSi2 thin-film heater and that using a target made from a mixture of $MoSi_2$ and Si powders with a molar composition ratio of Mo:Si = 1:X is called MoSiX thin-film heater.

The sputtering conditions were shown in a Table I. MoSi2 thin films fabricated by sputtering had very different orientations when the substrate was heated to a high temperature and when it was not heated[2]. Therefore, in this study, substrates were heated at 700 °C. At this temperature, highly oriented thin films were deposited with columnar structures[3]. The crystal structure of the thin films was analyzed by XRD. Composition analysis of the thin films was carried out using an SEM (JSM7500FA, JEOL) equipped with an EDS (EX-64195JMU, JEOL) at an acceleration voltage of 20 kV, a scanning area of 40×30 μm², a beam current of 200 pA, a sampling time of 300 s, and a count rate of 1100 cps. Mo and oxygen atoms were analyzed using the $L\alpha$ line and Si was analyzed using $K\alpha$ lines. A BSE detector (SM-34110, JEOL) was used with the SEM to analyze the composition on the surface of the grain boundary. A thin film for obtaining a BSE image was deposited at 700 °C for 10 min on an quartz substrate (diameter: 13 mm, thickness: 1 mm). The substrate was optically polished because the contrast of the BSE image depends on both the atomic weights of the elements on the surface and the surface flatness of the sample. The resistivity of the MoSiX thin film was measured by the four-probe method.

A fabricated MoSiX thin-film heater is shown in figure 1. A Pt electrode was deposited on the MoSi2 thin film by RF magnetron sputtering. The substrate temperature was 400 °C and the discharge time was 1 h. The distance between the two electrodes was 40 mm. A natural mica sheet (50×50 mm², thickness: approximately 0.2 mm) was inserted between a 5-mm-thick aluminum plate (used as a heat load) and the thin film deposited on a Si_3N_4 substrate to electrically insulate them. A constant dc voltage was supplied to the fabricated thin-film heater and its long-term stability was

Table I. Sputtering conditions.

discharge frequency	13.56 MHz
discharge power	200 W
discharge gas pressure	0.53 Pa
discharge gas	Ar
Ar gas flow rate	400 ml/min
substrate temperature	700 °C
discharge time	4 h

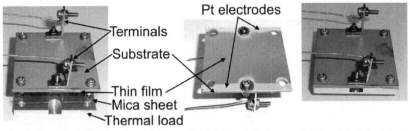

(a) Composition of heater   (b) thin film on substrate (c) heater with heat load

Figure 1. Structure of MoSiX thin-film heater.

evaluated from the changes in the resistance of the heater, the electric power, and the heating temperature. The heating temperature was measured using an alumel–chromel thermocouple inserted into a hole on the side of the aluminum plate.

RESULTS AND DISCUSSION

Evaluation of thin films

      XRD patterns revealed that the MoSiX thin-film consisted of a mixture of both Si with a cubic structure and $MoSi_2$ with a hexagonal structure (see figure 2). Furthermore, SEM images and XRD patterns of thin films showed that the MoSiX thin film deposited on the alumina substrate has a

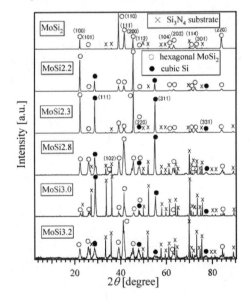

Figure 2. XRD patterns of MoSiX thin film.

Figure 3. SEM images of cross-section of MoSi2 thin film deposited on alumina substrate before heating.

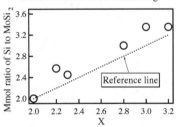

Figure 4. Molar ratio of Si to $MoSi_2$ for MoSiX thin film.

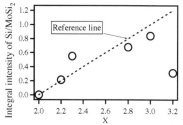

Figure 5. Ratio of the integrated intensity of
XRD peaks between $2\theta = 3$ and 90 ° for Si to
that for MoSi₂ for MoSiX thin film.

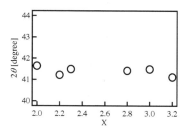

Figure 6. XRD peak angles for (100) plane
of MoSi₂ for MoSiX thin film.

columnar structure as shown in figure 3. All the thin films were approximately 10 μm thick. The Si and MoSi₂ particle distributions in the thin films could not be determined from elemental mapping based on EDS. The molar ratio of MoSi₂ and Si for the MoSiX thin film deposited on the Si₃N₄ substrate was measured by EDS (see figure 4). When the rate (X-2) of Si addition was increased, the measured molar ratio increased proportionally with X. This confirms that the added Si was incorporated in the thin film. Figure 5 shows the ratio of the integrated intensity of all the X-ray diffraction peaks between $2\theta = 3$ and 90 ° for Si to that for MoSi₂. This ratio increased proportionally with increasing X except for X = 3.2, although each atom had a different atomic scattering factor. The diffraction peak angles for the (100) plane of MoSi₂ were almost constant with X (see figure 6). However, no bulk compound having a higher Si content than MoSi₂ content has been discovered. Therefore, it is qualitatively speculated that the added Si is present in the thin film as a mixture with MoSi₂. However, the diffraction peak for MoSi₂ became broad and the substrate diffraction peaks were observed for X above approximately 2.8 (see figure 2), despite the thin film thickness being almost the same. The crystallinity of the thin film deteriorated at high X values. It is conjectured that Si atoms may enter MoSi₂ particles in the thin film and that a thin film of a compound that that does not exist as a bulk substance was produced by sputtering because Mo is a multivalent atom and the filling factor of an atom of hexagonal-type MoSi₂ is approximately 30 %.

Figures 7(a) and (b) show a BSE image and a concavo-convex image (TOPO image) of the MoSi3.2 thin film, respectively. The contrast of the spot shaped

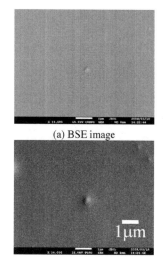

(a) BSE image

(b) TOPO image

Figure 7. (a) BSE and (b) TOPO images of
MoSi3.2 thin film on quartz substrate.

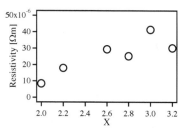

Figure 8. Resistivity of MoSiX thin film.

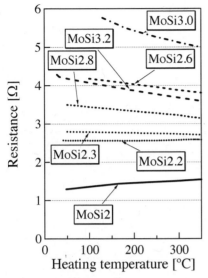

Figure 9. Resistance-heating temperature characteristic of each MoSiX thin-film heater.

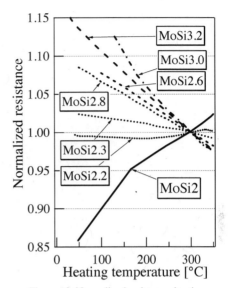

Figure 10. Normalized resistance-heating temperature characteristics of each MoSiX thin-film heater.

in the BSE image is not caused by a difference in the composition but by the unevenness of the thin film. Therefore, the formation of a microstructure originating from MoSi$_2$ and Si particles could not be observed in a BSE image with a magnification of 14,000.

Heating characteristics of MoSiX thin-film heaters

Figure 8 shows the resistivity of MoSiX thin films. The resistivity increased with increasing X due to the addition of Si, which has a higher resistivity than MoSi$_2$. Figure 9 shows the heating temperature characteristics of the resistance of the MoSiX thin-film heater. Figure 10 shows the heating temperature characteristics of the resistance of the MoSiX thin-film heater normalized by the resistance at 300 °C. Figure 11 shows the temperature coefficient α of the resistance. It was positive for X below approximately 2.2 and negative for X above approximately 2.2 and was almost constant for heating temperatures between room temperature and 350 °C. The temperature coefficient became almost zero for X of approximately 2.2 (see figures 9 to 11). In particular, the temperature coefficient was approximately -3.56×10^{-6} 1/°C in the temperature range between room temperature and 200 °C and was smaller than that (-5 to -10×10^{-6} 1/°C around room temperature) of manganin, which is used as a standard resistor.

Constant dc voltages of 12, 15, or 17 V were supplied to MoSi2, MoSi2.2, and MoSi3.2

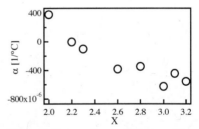

Figure 11. Temperature coefficient α of the resistance of MoSiX thin-film heater.

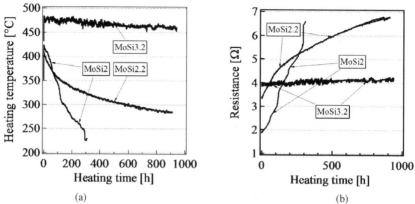

(a)   (b)

Figure 12. Heating-time dependence of heating temperature and resistance for MoSi2, MoSi2.2, and MoSi3.2 thin-film heaters.

(a)   (b)

Figure 13. SEM images of cross-section of (a) MoSi$_2$ and (b) MoSi3.2 thin films after heating test deposited on Si$_3$N$_4$ substrate.

thin-film heaters in air, respectively. The heating temperature and the resistance of the MoSi2 thin-film heater during the heating test were measured for 300 h from a starting temperature of approximately 420 °C, those of the MoSi2.2 thin-film heater were measured for 900 h from approximately 415 °C, and those of the MoSi3.2 thin-film heater were measured for 950 h from approximately 480 °C (see figure 12). For the MoSi2 thin-film heater, the heating temperature reached approximately 420 °C at an electric power of 75 W and its initial resistance in the heating test was approximately 1.9 Ω. The resistance increased with increasing heating time. The final resistance reached approximately 6.5 Ω and the heating temperature decreased to approximately 220 °C due to the increase in the resistance. For the MoSi2.2 thin-film heater, the heating temperature reached approximately 410 °C at an electric power of 65 W and its initial resistance in the heating test was approximately 3.3 Ω. Its resistance increased with increasing heating time. The final resistance reached approximately 6.6 Ω and the heating temperature decreased to approximately 290 °C. For the MoSi3.2 thin-film heater, the heating temperature reached approximately 480 °C at an electric power of 73 W and its initial resistance in the heating test was approximately 3.9 Ω. Its resistance became almost constant over 600 h and its final resistance was approximately 4.2 Ω and the heating temperature decreased to approximately 460 °C.

Thus, the MoSi3.2 thin-film heater was the most oxidation resistant and it operated stably in air over a long time at temperatures around 480 °C compared with the MoSi$_2$ thin-film heater. The other MoSiX thin-film heaters with X above 2.2 showed superior oxidation resistant than the MoSi2 thin-film heater. Thus, it was found that the oxidation resistance can be increased by adjusting the ratio of Si to MoSi$_2$ in accordance with the desired operating temperature.

Figure 13(a) shows a photograph of the cross-section of the MoSi2 thin film after the heating test. The initial color of the MoSi2 thin film was glossy gray, but after the heating test it had changed to blue. This blue is considered to be molybdenum blue which is a color peculiar to molybdenum oxide. Figure 14 shows XRD patterns of the MoSi2 thin film before and after the heating test. Diffraction peaks for MoO$_3$ were detected after the heating test[5]. It is speculated that the MoSi2 thin film oxidizes at high temperatures according to the following reaction:

$$2\,MoSi_2 + 7\,O_2 = 2\,MoO_3 + 4\,SiO_2,\quad (1)$$

or

$$2\,MoSi_2 + 3\,O_2 = 2\,MoO_3 + 4\,Si.\quad (2)$$

Diffraction peaks for Si were detected after the heating test[5]. However, SiO$_2$ is generated in very small quantities in air around 400 °C and SiO$_2$ in amorphous form may not give well defined diffraction peak but just very broad one. Therefore, it is unknown which equation is a main reaction type at a present stage. The increase in resistance is due to the formation of MoO$_3$, which is principally an insulator. Figure 14 shows that the MoSi2 thin film is oriented in the direction of the b-axis, since (110) and (021) diffractions were observed. Table II shows the composition analysis results for the cross-section of the MoSi2 thin film deposited on the Si$_3$N$_4$ substrate obtained using an EDS before and after the heating test. The composition was analyzed at the center of the thin films (indicated by the square A or B in figure 13(a)). The SEM images of the cross-section of the MoSi2 thin film shown in figure 13(a) reveal that another layer (the upper

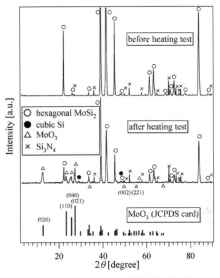

Figure 14. XRD patterns of MoSi2 thin film before and after heating test.

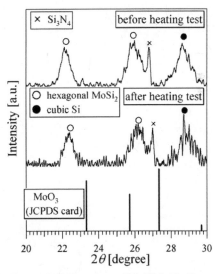

Figure 15. XRD patterns of MoSi3.2 thin film before and after heating test.

Table II. Composition ratios of Mo, Si, and O for MoSi2 and MoSi3.2 thin-film heaters before and after heating test. (See figure 13 about A and B)

	Composition ratio [mol%]		
	Mo	Si	O
MoSi2 before heating	33.9	66.1	
MoSi2 A: after heating	7.0	19.1	73.9
MoSi2 B:after heating	26.9	56.3	16.8
MoSi3.2 after heating	17.2	67.9	14.9

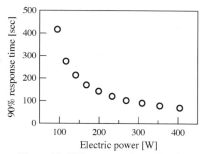

Figure 16. 90% response time–electric power characteristics of MoSi3.2 thin-film heater from room temperature to 400 °C.

layer, the square A) was formed on the thin film on the opposite side to the substrate. The columnar structure disappeared in the upper layer after the heating test, whereas, the columnar structure was preserved in the bottom layer (the square B). The oxygen content was over 70% in the upper layer and approximately 17% in the bottom layer. It is unknown why the oxygen content in the bottom layer becomes high after heating at the present stage. The resistance of the MoSi2 thin-film heater consists of the resistance of the thin-film and the contact resistance between Pt electrode and MoSi2 thin film. Pt silicate is formed around the Pt electrodes by high temperature heating above approximately 800 °C and is hardly formed at the temperature below 500 °C[2]. Furthermore, Pt silicate has the negative temperature coefficient of resistance and at high temperature, the resistance becomes small. Therefore, based on the SEM and EDS analysis results, it is speculated that the increase in the resistance of the MoSi2 thin-film heater is principally caused by an increase in the resistance of the upper layer of the MoSi2 thin film due to the generation of MoO$_3$. It is unknown why the concentration of silicon becomes high after heating.

Figure 13(b) shows an SEM image of the cross-section of the MoSi3.2 thin film. The SEM images of the cross-section of the MoSi3.2 thin film and the XRD patterns shown in figure 15 confirm that the columnar structure was preserved in the thin film after the heating test and an oxidized layer was not observed. The composition analysis results shown Table II revealed that the MoSi3.2 thin film had an oxygen atom content of approximately 15%. Despite the long time and the high temperatures (almost 480 °C) of the heating test, oxidation of the surface of the thin film could be prevented. Other thin films with composition ratio X also exhibited excellent oxidation resistances, similar to that of the MoSi3.2 thin film. Thus, it is concluded that adding an appropriate amount of Si to MoSi$_2$ is effective for improving its oxidation resistance.

The 90% response time of the heating temperature of the MoSi3.2 thin-film heater from room temperature to 400 °C was approximately 70 s at an electric power of 400 W (see figure 16). Even when an electric power of over 400 W was applied, deterioration (such as exfoliation of the thin film) was not observed. For all heaters, the resistance varied linearly with the heating temperature (see figures 9 and 10).

CONCLUSIONS

(1) The resistance of MoSi2 thin-film heaters used for a long time in air increased with increasing heating time at temperatures near 420 °C, because MoO$_3$, which is an insulator, was formed in the thin film due to the formation MoO$_3$ in the MoSi2 thin film.

(2) In the thin film fabricated by adding Si to $MoSi_2$ deposited on a Si_3N_4 or an alumina substrate by RF magnetron sputtering, no new compounds were generated except for $MoSi_2$ with a hexagonal structure and Si with a cubic structure.

(3) By adding Si to $MoSi_2$, the oxidation inside the thin film did not proceed in air at high temperatures around 480 °C. Thus, thin film heaters having excellent oxidation resistances could be fabricated.

(4) The formation of a microstructures due to $MoSi_2$ and Si particles could not be observed in back-scattered electron images. It is conjectured that Si atoms may enter $MoSi_2$ particles in the thin film and that a thin film of a compound that does not exist as a bulk substance was produced by sputtering.

REFERENCES

[1]K. Nakagawa, T. Tamamizu, N. Umeda, and T. Kawanami, *Fine Ceramics*, Giken, 100 (1982)
[2]Y. Ito, K. Wakisaka, M. Sato, and S. Yoshikado, *Key Engineering Materials*, **320**, 95–98 (2006)
[3]S. Hayashi and S. Yoshikado, *Key Engineering Materials*, **216**, 105–108 (2002)
[4]K. S. Sree Harsha, *Principle of Physical Vapor Deposition of Thin Film*, Elsevier, 535 (2006)
[5]L. G. Berry and J. V. Smith, *Powder Diffraction File*, American Society for Testing and Materials, 5-508, 5-565

INFLUENCE OF INTERFACE ON TUNABILITY IN BARIUM STRONTIUM TITANATE

Naohiro Horiuchi, Takuya Hoshina, Hiroaki Takeda, Takaaki Tsurumi
Graduate School of Science and Engineering, Tokyo Institute of Technology
Ookayama, Meguro, Tokyo, Japan

ABSTRACT

The influence of interface layer on a tunability of parallel plate $(Ba, Sr)TiO_3$ thin film capacitors was investigated. BST thin film capacitors with top electrode of Pt, Au and Ag were fabricated. BST films had thickness of 40, 60, 80 and 120nm. Apparent dielectric constants and tunability were depending on the work function of top electrode metal. The tunability increased with increasing the BST film thickness. The interfaces between BST films and electrodes were considered as Schottky junctions. Thus, depletion layers were formed in the interfaces depending on the difference of the work function of metal electrodes. Capacitance-voltage characteristics showed that the increase of the depletion layer thickness provides low tunability. The applied voltage to the depletion layer does not contribute to lowering the capacitance.

INTRODUCTION

$BaSrTiO_3$ (BST) thin films have been intensively studied for the application of microwave tunable devices.[1,2] Several studies have indicated that permittivity decreased with decreasing BST film thickness and were dependent on metal of electrodes.[3-6] These phenomena were attributed to existences of interface layers having different dielectric properties from those of the interior BST films.[7-10]

A Schottky barrier is formed at the interface between a metal and an oxide ferroelectric material such as barium strontium titanate.[10,11] Electric field is induced in a depletion layer, the interface layer of Schottky junction. In BST films, dielectric permittivity firmly depends on electric field due to the nonlinear dielectric property of the BST. The permittivity in the depletion layer must be depressed by the electric field. This low permittivity interface lowers the capacitance of BST thin film capacitors.[12] The reason why the permittivity is dependent on the film thickness and electrode metal is existence of this interface layer.

The existence of the depletion layer may affect tunability of thin film capacitors. In this case, effects of the external and internal electric field are superposed. In depletion layers, both internal bias voltage induced by Schottky barrier and external bias voltage are applied. This complexity makes it hard to understand the influence of Schottky barriers on the BST thin film capacitors. To visualize the influence of the depletion layers, we fabricated thin film BST capacitors with different kind of metal electrodes, in order to obtain different thickness of depletion layers, and measured these Capacitance-Voltage characteristics.

BACKGROUND

Figure1 (a) and (b) show an energy band diagram of a p-type semiconductor (BST)[11] and metal that making a Schottky junction. A potential barrier and a depletion layer are formed at the interface owing to the difference in work functions between BST and electrode metal. The thickness of the depletion layer is given by, [13]

$$t_d = \sqrt{\frac{2\varepsilon_0\varepsilon_d}{eN} V_{bi}} \quad , \qquad (1)$$

where ε_0 is the dielectric constant of vacuum, ε_d is the dielectric constant of the depletion layer, e is the elementary electric charge, N is the carrier concentration in the BST films and V_{bi} is the potential barrier height at the interface. In this formula, variation of permittivity ε_d is neglected. Permittivity must depend on electric field in the depletion layer.

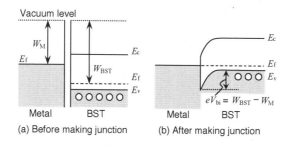

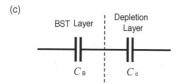

Fig.1: (a), (b) Energy band diagram before and after making junction of a metal and BST film (p-type semiconductor). E_f: Fermi level, E_c: bottom of conduction band, E_v: top of valence band, W_M: work function of a metal, W_{BST} : work function of BST film, V_{bi} : potential-barrier height. (c) An equivalent circuit of BST film. C_d and C_B are capacitances of the depletion layer and interior BST layer, respectively.

Figure1(c) shows an equivalent circuit of BST film. The depletion layer is connected with interior BST layer in series. An interface layer between BST and SRO film is ignored because

BST/SRO junction did not form depletion layers.[5,9] When external bias voltage V is applied to the BST film, the voltage applied to the depletion layer V_d is calculated by,

$$V_d = \frac{C_B}{C_d + C_B} V, \qquad (2)$$

where C_d and C_B are the capacitances of the depletion layer and interior BST layer, respectively. The smaller C_d provides the larger V_d. This means that large voltage is applied to the depletion layer when the capacitance of the depletion layer is small. And wider depletion layers have high voltage.

EXPERIMENT

BST thin films were prepared with an RF magnetron sputtering on $SrRuO_3$ (SRO) / [100] - $SrTiO_3$ (STO) single crystal substrates. The SRO films were deposited as bottom electrodes using the same sputtering system. The BST target is stoichiometric $Ba_{0.5}Sr_{0.5}TiO_3$. The sputter gas was Ar and O_2 mixture (Ar/ O_2 = 4:1). The gas flow rates were adjusted for a total pressure of 30 mTorr. The thicknesses of the BST films were 40, 60, 80 and 120 nm. Top electrodes metals Pt, Au, and Ag were deposited on the BST film with a DC sputtering. Top electrodes were processed into circle patterns (diameter =80 μm) by using a lift-off method. Dielectric properties were measured with an impedance analyzer (HP 4294A) at room temperature.

RESULT AND DISCUSSION

Figure 2 shows the inverse capacitance density (A/C) as a function of film thickness. The capacitance values are at 10 kHz. The data could be described using a model of the depletion layer that connected in series with an interior BST film of permittivity ε_B :

$$\frac{A}{C} = \frac{t - t_d}{\varepsilon_0 \varepsilon_B} + \frac{t_d}{\varepsilon_0 \varepsilon_d}, \qquad (3)$$

where A, C, t are the capacitance area, capacitance and total BST film thickness, respectively. Here, permittivity of the depletion layer is far less than that of interior BST film and probably does not depend on the metal electrodes. The intercept at $t=0$ corresponds to a capacitance of depletion layer. We can evaluate the depletion layer thickness for each kind of metal electrode. Thus, the intercepts indicate that thickness of the depletion layer between Ag and BST is the largest, subsequently Au and Pt. Schottky barrier heights are connected with Depletion layer thicknesses by equation (1). The order of Schottky barrier height (Ag>Au>Pt) obtained from the intercepts is reverse order of work function (Ag: 4.60 eV, Au: 5.38 eV, Pt: 5.63 eV). These results are consistent with the formation of Schottky junction between p-type semiconductor (BST) and metal.

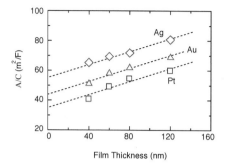

Fig.2: Inverse capacitance density as a function of BST film thickness measured at 10 kHz.

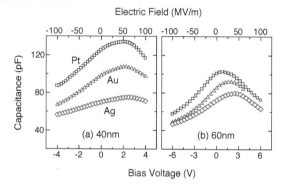

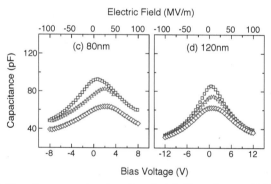

Fig 3: Capacitances as a function of applied dc bias voltage at 10 kHz for different metal electrode and BST film thickness.

Figure 3 (a)-(d) show Capacitances as a function of applied dc bias voltage at 10 kHz for each BST film thickness: (a) 40 nm, (a) 60 nm, (a) 80 nm and (a) 120 nm. Each figure has the results of different metal electrode Pt, Au and Ag. The positive direction of voltage is determined for the forward bias direction in Schottky junction. As shown on upper and lower abscissa axes, we applied different voltages in order to generate the same electric field on the films with different thickness. The maximum capacitances on the curves are not obtained at zero bias voltage, but at forward bias voltage. Furthermore, deviations from zero bias voltage are increasing with the thickness of the depletion layers. Figure 4 shows tunability as a function of film thickness. Tunability n is defined as below:

$$n = (C(0) - C(V))/C(0) \qquad (4)$$

where $C(0)$ and $C(V)$ are the capacitance at zero and V bias voltage. In figure 4, the values of V are the voltage corresponding to electric field 100MV/m. Tunability increase with increasing film thickness in each kind of metal electrodes. In addition, in all the BST film thickness, the tunability become lower in order of Pt, Au and Ag.

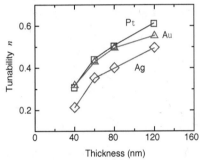

Fig 4: Tunability as a function of film thickness for each metal electrode.

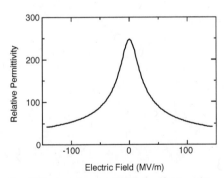

Fig 5: Tunability curve calculated from the model of Landau-Devonshire-Ginzburg.

This order can be explain by the thickness of the depletion layers. The thinner depletion layer provides the smaller capacitance (see eq.3). So, external bias voltage applied to the depletion layer is increasing with increasing the depletion layer thickness. On the other hand, the voltage applied to interior BST film is decreasing. Figure 5 shows tunability curve calculated from the model of Landau-Devonshire-Ginzburg. The permittivity under low electric field decreases steeply. By contrast, the permittivity in the high electric field is an almost constant. This expects that the permittivity of the depletion layer decreases much less than that of interior BST layer decreases, because the depletion layer previously has high internal electric field. Thus, external bias voltage applied to the depletion layer is the reason for lowering tunability. When the total BST film thickness increases, CB becomes smaller. So, the voltage applied to the depletion layer decreases. Therefore, tunability increases with increasing film thickness.

CONCLUSION

Tunability was influenced by the kind of metal electrodes because the thickness of the depletion layer varies with the work function of the metal electrodes. The increase of the thickness of the depletion layer provides low tunability. Because external bias voltage applied to the depletion layer increases with increasing the thickness of the depletion layer, and voltage applied to the interior BST film decreases. The applied voltage to the depletion layer does not contribute to lowering the capacitance, because tunability of the depletion layer is much lower than the interior BST film.

REFERENCES

[1] P. Bao, T. J. Jackson, X. Wang and M. J. Lancaster, *J. Phys. D: Appl. Phys.*, **41**, 063001 (2008).

[2] A. K. Tagantsev, V. O. Sherman, K. F. Astafiev, J. Venkatesh, and N. Setter, *J. Electroceram.*, **11**, 5 (2003).

[3] M. S. Tsai, S. C. Sun, and T. Y. Tseng, *IEEE Trans. Electron Devices.*, **46**, 1829 (1999).

[4] H. Seokmin, B Heungjin, A. Ilsin, and K. Kyung, *Jpn. J. Appl. Phys., Part 1*, **39**, 1796 (2000).

[5] B. T. Lee and C. S. Hwang, *Appl. Phys. Lett.*, **77**, 124 (2000).

[6] U. Ellerkmann, R. Liedtke, U. Boettger, and R. Waser, *Appl. Phys. Lett.*, **85**, 4708 (2004).

[7] D. S. Boesch, J. Son, J. M. LeBeau, J. Cagnon, and S. Stemmer, *Appl. Phys. Exp.*, **1**, 091602 (2008)

[8] N. H. Finstorm, J. Cagnon, and S. Stemmer, *J. Appl. Phys.*, **101**, 034109 (2007).

[9] L. J. Sinnamon, R. M. Bowman, and J. M. Gregg, *Appl. Phys. Lett.*, **78**, 1724 (2001).

[10] C. S. Hwang, B. T. Lee, C. S. Kang, K. H. Lee, H. J. Cho, H. Hideki, W. D. Kim, S. I. Lee, and M. Y. Lee, *J. Appl. Phys.*, **85**, 287 (1999).

[11] J. F. Scott, *Integr. Ferroelectr.*, **9**, 1 (1995).

[12] N. Horiuchi, T. Matsuo, T. Hoshina, H. Kakemoto and T. Tsurumi, *Appl. Phys. Lett.*, **94**, 102904 (2009).

[13] S. M. Sze, *Physics of Semiconductor Device*, 2nd ed. (wiley, New York, 1981).

RECENT PROGRESS IN MULTILAYER CERAMIC DEVICES

Hiroshi Takagi

Murata Manufacturing Co., Ltd.
10-1, Higashikotari, 1-chome, Nagaokakyo-shi, Kyoto 617-8555 Japan

Abstract
Recent advances in multilayer ceramic devices using in electronic circuits are presented. The multilayer devices are including MLCC, LTCC and PTC thermistors. Their processes controlling chemical compositions and crystal structures of materials for bringing out the suitable functions and the electric properties are explained. These new multilayer devices will apply in the various electronic circuits, which the manufacturers of the electronic equipments demand to realize at present or in future.

1. Introduction

Murata manufactures many kinds of multilayer ceramic devices. We started to make MLCC in 1970. After that we released many kinds of multilayer ceramic devices like LTCC, ferrite beads, chip coil, piezoelectric devices, PTC thermistor and so on.

Manufacturing technologies using in these multilayer devices have much in common, on the contrary, their material design has each specific issues. Their processes include controlling chemical compositions and crystal structure of materials, forming and stacking green sheets, firing them to rigid bodies and applying electrodes for bringing out the electric properties. Realizing multilayer devices suitable for functionally, these processes have to be highly sophisticated ones. At the same time, structural and circuit design technology is very important.

In this paper, I would like to introduce Murata's resent MLCC, LTCC and PTC thernistor using multilayer technology. Especially their material design and processing will be presented.

2. MLCC

Recently, multilayer ceramic capacitors (MLCC) using the nickel-electrodes intend to have higher performances such as miniaturizing, larger capacitance, high heat-resistance and so on. Here, I will present the MLCC with low microphonics [1] and the MLCC with large capacitance [2].

2.1 MLCC with low microphonics

The manufacturers of electronic equipments and the module demand low distortion lately to improve the sound quality from capacitors used for decoupling in sound input circuits. In the display market, customers demand no microphonics from capacitors used for Liquid Crystal Display (LCD) and Plasma Display Panel (PDP), because no sound noise is tolerated while in operation.

The common MLCCs of Class2 achieve a high dielectric constant and consequently high capacitance. But these capacitors have high distortion and large microphonics, so these properties make them unsuitable for the applications. We have successfully developed a new MLCC with both high capacitance and low microphonics by using ceramic materials with controlled dielectric characteristics.

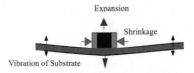

Fig 1. Microphonic generation mechanism.

The microphonics generation mechanism is illustrated in Fig 1. For a MLCC, this expansion and contraction occurs in the direction of the chip thickness. The plane of the chip parallel to a mounted substrate contracts and expands in the reverse direction. This displacement is transmitted to the substrate, and the substrate vibrates in the direction of the thickness. When the amplitude is amplified by the substrate, the frequency in the audible range (50Hz~20kHz), is perceived as sound.

We have succeeded in developing a new ceramic material with low dielectric loss, distortion and microphonics [1]. This ceramic material was used to develop capacitors between 0.1~1.0μF/25~ 100V with very low distortion and microphonics. This development would allows MLCCs to replace other types of capacitors currently used in this market segment. Figure 2 shows the D-E hysteresis curves and the distortions in the electric field of two materials. The new dielectric material does not have the spontaneous polarization and the hysteresis curve is para-electric in nature. As for the new one, the distortion in the electric field dropped to 1/8 of that of the traditional material. Figure 3 shows the sound pressure of materials. As a result, the microphonics (sound pressure) of the new product is significantly lower.

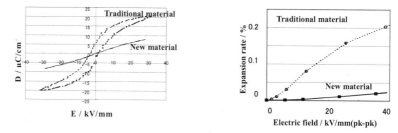

Fig 2 D-E hysteresis curves and the distortions in the electric field.

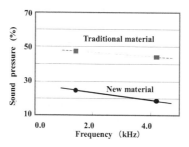

Fig.3   Frequency dependence of sound pressure.

2.2 MLCC with large capasitance

These days, MLCs with 1μm dielectric layer in thickness have already appeared on the world market. The market for large capacitance capacitors, more than 10 μF, is recently expanding and replacing that for Ta and Al electrolytic capacitors. MLCs with large capacitance commonly consist of BaTiO$_3$ based ferroelectric material for dielectrics and Ni metal for inner electrodes, so MLCs have to be fired in a reducing atmosphere to prevent Ni electrodes from the oxidation. The core-shell BT ceramics are widely used for MLCs with the specification of X7R (EIA code: ΔC/C=±15 % at -55 °C to 125 °C), showing a stable temperature dependence of the dielectric constant and a high reliability. As shown in Fig. 4(a), the core part of the core-shell BaTiO$_3$ is pure BaTiO$_3$, which is characterized as the ferroelectric phase, and the shell part, which was modified with some dopants, is the paraelectric phase or the diffused ferroelectric phase at room temperature. The temperature dependence of the dielectric constant is considered to be a combination between dielectric constants of the core part and the shell part. The high reliability is considered to be due to the high reliability of the shell, which controls the composition of the shell part.

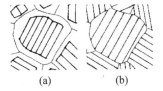

(a)   (b)

Fig 4. Schematic diagrams of the core-shell structure (a) and the non-core-shell structure (b). Solid lines in grains mean the ferroelectric domain structure.

Fine-grained dielectric ceramics with high dielectric constant have been required for thinner dielectric layers. However, it is well known that ferroelectric materials like BaTiO$_3$ ceramics have some degree of grain size dependence in ferroelectricity and the dielectric constant decreases depending on the decrease of grain diameter. The core-shell ceramics were considered to be especially difficult to possess high dielectric constant in very fine grain ceramics, because of smaller size of the core part possessing ferroelectricity. Instead of the core-shell dielectrics, alternative fine-grained ceramics without the core-shell structure, as shown in Fig. 4(b), shows a high dielectric constant, because grains show high ferroelectricity throughout the entire volume of the grains, and a stable temperature dependence of the dielectric constant similar to the core-shell dielectric ceramics. The reason why they have the stable temperature dependence is assumed as follows. In the BaTiO$_3$ ceramics, the grain growth is inhibited during sintering as in the core-shell ceramics. The inhibition of the grain growth results in large stress in the grains from grain boundaries and hence the maximum dielectric constant at the Curie temperature is depressed by the surface stress of grains. The high reliability of the new dielectrics without the core-shell structure results from the high reliability of grain boundaries enforced with some additional elements.

We use Ca-doped BaTiO$_3$ as a dielectric ceramics without the core-shell structure [2]. Figure 5 shows temperature dependences of dielectric constant of the Ca-doped BaTiO$_3$ ceramics, revealing the difference between the ceramics with grain growth and without grain growth during sintering. The Ca-doped BaTiO$_3$ ceramics with grain growth show typical temperature dependence of dielectric constant with a high dielectric constant at the Curie temperature. On the other hand, the ceramics without grain growth show stable temperature dependence, with the dielectric peak at the Curie temperature depressed and showing a high dielectric constant over wide temperature range. There is no need to obtain stable temperature dependence to have the core-shell structure.

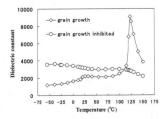

Fig 5. Temperature dependence of dielectric constant of Ca-doped BaTiO$_3$ ceramics.

Fig. 6 is the TEM image of the Ca-doped BaTiO$_3$ ceramics without grain growth, showing that the grain boundaries were at most 1 nm width. This means the Ca-doped BaTiO$_3$ ceramics without the core-shell structure is realized. Figure 7 shows 113 and 311 diffractions of the calcined powder, the surface of the sintered ceramics and the surface of the chemically etched ceramics. The sintered ceramics show a broad peak characterized as having a large non-uniform distortion in the lattice in comparison with the calcined powder. By etching grain boundaries, it is confirmed that the tetragonality of the Ca-doped BaTiO$_3$ ceramics recovers to that of the calcined original Ca-doped BaTiO$_3$ powder. This means that grain boundaries in the ceramics without grain growth have caused a large stress in the grains, resulting non-uniform distortion. XRD data of the Ca-doped BaTiO$_3$ ceramics was similar to that of the calcined powders. The stable temperature dependence in dielectric constant is considered to relate to the existence of grain boundaries yielding a huge stress. The typical cross-section of MLCC with 1μm thickness of dielectric layer using Ca-doped BaTiO$_3$ ceramics is shown in Fig 8.

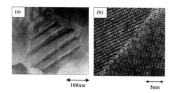

Fig 6. TEM image (a) and TEM lattice image of the Ca-doped BaTiO$_3$ ceramics with the grain growth inhibited.

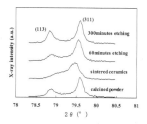

Fig 7. XRD patterns of Ca-doped BaTiO$_3$ ceramics and calcined powder.

Fig 8. Cross-section of MLCC with 1μm dielectric layer in thickness.

3. LTCC

In the field of wireless communication, the LTCC substrates are expected as one of the most promising technology to realize the miniaturization of RF circuits. LTCC is applied to two major applications. One is a chip-type functional devise such as the band pass filters, duplexers, baluns and couplers. Another is a large-size circuit board used for the RF modules of mobile phones like the Bluetooth modules, automotive electrical control units and so on, on which are composed of the active devices mounted on the LTCC substrates and the embedded passives in the substrates such as strip line, coil, resonator, capacitor and ground plane.

However, there are expanding demand for further miniaturization and integration of the RF circuits. Therefore co-firing technology of two or more dielectric materials and resistive materials will be developed further to integrate capacitors and resistors in substrate. At the same time, lines and space of electrode should be finer. On the other hand, cost-down is always the challenge of LTCC. Large-sizing of substrate has the advantage for cost-down because the process cost can be decreased. Besides thin substrate has the advantage for cost-down and downsizing of substrate. But for these requirements, mechanical strength of LTCC substrate will be increased. In addition, the wireless applications have broadened from the cellular phones at low microwave frequency towards the applications at millimeter wave region such as 30 GHz and beyond, then, LTCC of higher Q value will be needed further.

In this section, we focus on trends of the material and process designs of novel LTCC systems for new wireless applications including the low loss dielectric materials at higher frequencies, the co-firing technology of two materials with different dielectric constants, the embedded passives such as buried resistors and non-shrinkage firing method [3].

3.1 High-Q LTCC material

To meet the requirements to have ever-higher processing speed and higher integration density, LTCC materials are being used at microwave frequencies. But the conventional ceramic materials share a common problem: their Q values and mechanical strength are lower than those of crystal ceramics. LTCC materials mainly consist of ceramic filler and glass, and these components reduce the Q value and mechanical strength. In particular, for very high frequency applications, the Q value is a more important property for LTCC materials than the other properties. In response to this problem, LTCC materials called NGC (product's name) that have a high Q value at millimeter wave frequencies have been developed. Table 1 shows the characteristics of NGC. The mechanical strength and thermal conductivity values of NGC are higher than those of conventional LTCC materials; this is due to the high crystallization of NGC. Figure 9 shows the XRD pattern of NGC. The main crystal phases were those of $MgAl_2O_4$ (Q at 14GHz; 8400), $Mg_3B_2O_6$ (Q at 16GHz; 9400) and Li_2MgSiO_4; small amounts of $Mg_2B_2O_5$ and Mg_2SiO_4 crystal phases were also detected. No halo indicating the presence of glass phases were found in the XRD pattern, leading us to conclude that there are very little glass phases that have a low Q value.

Table 1   Characteristics of NGC.

Qf (GHz)	53000
	@24GHz
TCC (ppm·K^{-1})	+170
Flexural strength (MPa)	250
Thermal conductivity (W·m^{-1}·K^{-1})	5.59
Termal expansion (K^{-1})	11×10^{-6}
Cofireable electrode	Cu

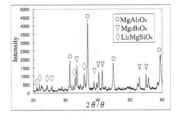

Fig. 9   XRD (Cu kα) chart of NGC

Figure 10 shows a cross-sectional view of devices co-fired with copper inner electrodes and via holes. The figure indicates that NGC has good compatibility with copper inner electrodes and via holes. The band-pass-filter (BPF) for 26GHz is shown in Fig.11 and its frequency characteristics on the test board is shown in Fig.12.

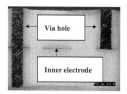

Fig.10   Cross-sectional view of NGC co-fired with copper inner electrodes and via holes

Fig.11   The BPF is connected by solder (3.2x2.5x1.2mm).

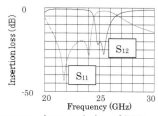

Fig.12   Frequency characteristics of BPF on the test board.

3.2 Novel LTCC with embedded LCR

We developed LTCC material called AWG, for non-shrinkage firing method. We selected the glass powder, in which β-wollastonite ($CaSiO_3$) is crystallized, because it is useful to increase its strength. Actually, we use the mixture of around 50 vol% Al_2O_3 and 50 vol% glass. The microstructure of AWG is shown in Fig.13.

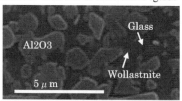

Fig. 13   SEM image of AWG.

We can also use constrained sintering technique for AWG. Constrained sintering is same meaning as "zero-shrink" or "non-shrinkage" processing. In this case, constrained sheets composed Al_2O_3 are stacked on the both sides of dielectric green sheets, and they are removed after sintering. Table 2 is the comparison between normal sintering with shrinking and constrained "non-shrinkage" sintering. By using constrained sintering technique, a shrinkage tolerance and linearity of 100mm x 100mm scale substrate are improved significantly, as shown in the table 2. The tolerance decreases to one-fifth, and the linearity improves to one-third, if we use the constrained sintering.

Table 2　Comparison between normal and constrained sintering
(100mm x 100mm substrate)

	Constrained	Normal
Shrinkage tolerance/ μm	±60	±300
Linearity/ μm	±50	±140
Electrode roughness, Rmax/ μ	<2.5	<3.1

In addition to above-mentioned process, a lot of process technologies are needed to manufacture highly integrated substrates. As shown in Figure 11, we have many process technologies for AWG subatrate, for example, formation of fine line, cavity and thermal via, formation of buried inductor, capacitor and resistor, and we can trim the buried elements using a laser-trimming technology. Ni/Au plates the surface conductors chemically.

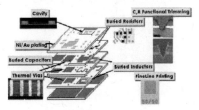

Fig.14　Various process technology for AWG substrate.

We use pure Ag for conductor of AWG. So its electrical conductivity is about 1.9 mΩcm, and the thickness of 150μm line is 14μm. The line and space of 50μm/50μm is available. Diameter of via is 100, 150, 200 or 300μm, and the pitch among them is 500μm for 150mmϕ via. We use RuO_2-base materials for surface and buried resistor. The range of sheet resistivity is 10Ω–10MΩ, and TCRs are +/-150ppm for surface one and +/- 200ppm for buried one. The minimum size of resistor is 0.25mm by 0.25mm. Trimming of buried resistor is available. Flexural strength is very strong, such as 300 MPa, It is suitable to realize thin substrate. To decrease a total height of devices, thin substrate is required. We are manufacturing the substrate of 200μm thickness.

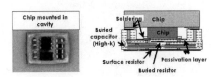

Fig.15　Model device using AWG substrate.

Figure 15 is model device for cellular phone including cavity structure, buried capacitor with high-ε_r materials, and buried resistor. As shown here, we can realize highly integrated electronic devices using our novel LTCC (AWG substrate).

Murata developed the LTCC system that consists of three co-firable materials. Three materials are AWG with low-ε_r (8.7), RWG with high-ε_r(15.1) and a resistor material with low temperature coefficient of resistance (TCR), which is considered to be a solution for further development of RF circuits. Such a system has several advantages., for instance, cross-talk noise between lines and electric-signal delay are suppressed by positioning electric wiring lines mainly on low-ε_r material layers. Internal capacitors can be downsized by forming them onto high-ε_r material layers. The resistors can be buried in the AWG/RWG and AWG/RWG interfaces. Using three kinds of resistive paste, we can form resistors from 10 to 1000Ω, on two interface positions without any defects like delamination. Buried resistors were trimmed using a YAG laser. No change in the quality of resistor materials was observed in resistance measurements or EDX.

Figure 16 is an SEM image of RWG material. Ba-(Sm, Nd)-Ti-O and Al$_2$O$_3$ crystals were dispersed in a glass matrix. Figure 17 is an XRD pattern of RWG. Al$_2$O$_3$ and CaSiO$_3$ (wollastonite) were detected as main phases. Al$_2$O$_3$, CaSiO$_3$, Ba-(Sm, Nd)-Ti-O, and CaTiO$_3$ were detected in RWG. CaTiO$_3$ is generated from the reaction of glass with Ba-(Sm, Nd)-Ti-O and has a highly negative TCC. On the other hand, the TCCs of glass, Al$_2$O$_3$ and CaSiO$_3$, are positive, so those ingredients in RWG precisely compensate each other for near-zero TCC of RWG.

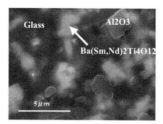

Fig.16   SEM image of RWG substrate.

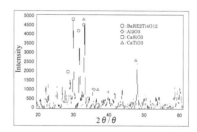

Fig.17   XRD diffraction pattern of RWG.

Figure 18 shows shrinkage curves of both materials. The start temperature and amount of shrinkage of both materials are almost the same. The reason for this seems to be that both materials contain common glass, and the difference in composition of both materials is small and only in filler contents.

Thermal expansion coefficients (TECs) of AWG and RWG are listed in Table 3. Ba-(Sm, Nd)-Ti-O has a high TEC (10ppm/°C), but RWG contains more glass (low TEC) than AWG and is highly crystallized (CaSiO$_3$; low TEC), so the deviations of TECs from AWG in ingredients of RWG cancel each other. As a result, RWG has the same TEC as AWG. So the TECs of both materials are well matched, and stress and delamination between co-fired materials is completely suppressed.

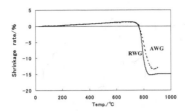

Figure 18 Shrinkage-curves of AWG and RWG substrates.

Table 3　Thermal expansion coefficients.

	TEC/ ppmK^{-1}
AWG	7.6
RWG	7.6

(600°C – Room Temperature)

Figure 19 shows two types of cross-section with asymmetric layer structure of co-fired materials. The AWG and RWG materials have the same TEC, so no delamination or warpage occurs. This allows various geometric structures of substrates to be designed. In this LTCC system, capacitors with high Q and near zero TCC can be miniaturized by RWG and low TCR resistor can be buried and trimmed by YAG-laser in substrate. A Ni/Au electrode can be plated on the surface of Ag electrode lines, as shown in Fig. 20, and no damage to the substrates occurs.

Fig.19　Cross-section of co-fired substrates AWG/RWG.

Fig.20　Image of Ni/Au-plated substrate (100x100 mm) using AWG/RWG.

The co-fired substrate is very stable chemically and mechanically. We believe that the LTCC substrate created by co-firing holds great promise for the development of functional substrates and module products to meet increasing needs for miniaturization given the added acceleration and expansion of usable frequencies of wireless communications equipment.

4. PTC thermistor

It is required to decrease the room-temperature resistivity of semiconducting BaTiO₃ ceramics in PTC thermistors (positive temperature coefficient resistor). Although there are many approaches to decrease the resistance of a PTC thermistor, it is difficult to further decrease the resistivity of BaTiO₃ ceramics. On the other hand, if a PTC element with a multilayer structure can be fabricated, it will have a very low resistance. However, it is difficult to co-fire BaTiO₃ with PTC characteristics and an internal electrode because of the following reasons. It is commonly known that donor-doped BaTiO₃ ceramics fired in air exhibit positive temperature coefficient of resistivity (PTCR) characteristics. Its resistivity increases drastically at the Curie temperature of BaTiO₃. The PTCR characteristics are reported to be brought about by the acceptor state at the grain boundary due to absorbed oxygen. However, the PTCR characteristics are substantially decreased when BaTiO₃ is fired in a reducing atmosphere because of the desorption of oxygen. However, as the inner electrodes should form an ohmic contact with the n-type semiconducting BaTiO₃, these electrode should be composed of a base metal such as Ni and be fired in a reducing atmosphere to prevent the oxidization of the base metal. Therfore, it is difficult to co-fire BaTiO₃ with PTCR characteristics and an internal electrode.

By significantly reconsidering the material composition and the process, we succeeded in the commercial production of the multilayered type of PTC thermistors for the first time in the world. Various technical issues that were faced until the commercial production of the thermistor and the features of the thermistor as the commodity of chip PTC thermistors are introduced below.

At the grain boundary, an electron is trapped along with the adsorption oxygen and acceptor element and a Schottky barrier is formed. The potential barrier ϕ is given by

$$\phi = \frac{eN_s^2}{8\varepsilon_r\varepsilon_0 N_d} \qquad (1)$$

where e denotes the charge of an electron; N_d, the concentration of the charge carriers; N_s, the density of surface states; and ε_r, the relative permittivity of the grain boundary region [1]. BaTiO₃ is ferroelectric at temperatures below Tc; therefore, this potential barrier height is low and exhibits a low resistance. On the other hand, because of the occurrence of paraelectric material beyond Tc, it exhibits PTC characteristics.

In Eq.(1), the density of surface states N_s, which decides the potential barrier height, originates in the adsorption oxygen [5–8]. The decrease in the oxygen partial pressure during firing leads to the deterioration of the PTC characteristics, as shown in Fig.21. Therefore, it is necessary to increase the oxygen partial pressure during firing in order to obtain satisfactory PTC characteristics.

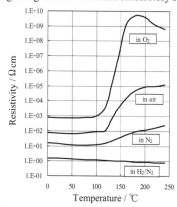

Fig. 21   Deterioration of PTC characteristics with decrease in oxygen partial pressure.

Incidentally, because BaTiO$_3$ is an *n*-type semiconductor, precious metals such as Pd that generate a Schottky barrier at the interface with the ceramic and give rise to a high resistance cannot be used as internal electrodes. On the other hand, when a base metal such as Ni is used, an ohmic contact is obtained and it exhibits a resistance characteristic of ceramics. Thus, it was difficult to satisfy the maintenance of the PTC characteristic and the oxidation prevention of a base metal internal electrode at the same time. By reconsidering the material composition and the process, we found some BaTiO$_3$ materials with excellent resistance to reducing atmosphere [4].

The withstand voltage of PTC thermistors is decided by the thermorunaway; therefore, a high withstand voltage is indicated so that the peak value of the resistance (substitute it by the resistance at 250°C, R_{250}) is generally high. However, the influence of the thickness of the ceramic layers is also large. This is because the Schottky barrier height of the grain boundary decreases when the voltage is applied, as shown in Fig.22. When the voltage is applied, the parameter ϕ in Eq.(1) decreases in a manner shown in the following equation.

$$\phi = \phi_0 - eV \qquad (2)$$

Here, ϕ_0 denotes the Schottky barrier height before applying the voltage; V, the voltage applied to the grain boundary; and e, the charge of an electron.

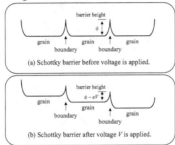

(a) Schottky barrier before voltage is applied.

(b) Schottky barrier after voltage V is applied.

Fig. 22　Decrease in Schottky barrier height when voltage is applied.

The R_{250} value was found to have decreased from 1 V or more, as shown in Fig.23, when the voltage dependency of the PTC characteristics was actually examined for a ceramic layer of thickness 20 μm. Although the voltage was only 1 V, an electric field of 50 V/mm was applied per unit length because the ceramic layer was thin; moreover, it is considered that the R_{250} value decreased substantially. In addition, the voltage was mainly applied to the grain boundary, and a considerably high electric field was definitely applied to the grain boundary. Only the grain boundaries of 3–4 piece exist in a ceramic layer of 20 μm thickness because it is approximately 5 μm even if the grain size is minimum in the conventional PTC thermistor. It is expected that a high electric field of 0.2 V (1000 V/mm) is applied to the extremely thin barrier layer.

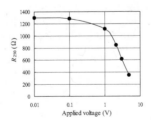

Fig. 23　Applied voltage dependency of R_{250} value.

Therefore, it is necessary to improve the resistance substantially by controlling the applied voltage per grain boundary, that is, by minimizing the grain size to maintain a high withstand voltage, even if the thickness of the ceramic layer is reduced.

The characteristics of the commercially produced multilayered chip PTC thermistors are shown in Fig.24. A resistance of 0.2 Ω was obtained for the size of 2012, and a low resistance of 1/100 was achieved as compared to that of the conventional bulk type of chip PTC thermistors. The non-operating current was improved by 500 mA from tens of mA. Moreover, a withstand voltage of 6 V was also achieved by minimizing the grain size, and a sufficient withstand voltage could be maintained for low-voltage circuits such as personal computer peripherals.

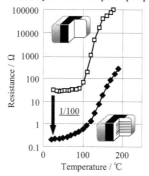

Fig. 24   Characteristics of chip PTC thermistor for overcurrent protection.

6. Summary

Recent advances in multilayer ceramic devices including MLCC, LTCC and PTC thermistors are presented. For the newly MLCs, the MLCC with low microphonics and the MLCC with large capacitance were explained. For the newly LTCCs, high-Q LTCC material and LTCC with embedded LCR were presented. Finally, PTC thermistor with multilayer structure was introduced. Especially, their processes controlling chemical compositions, crystal structure and firing profiles were discussed in detail. These new multilayer devices will apply in the various electronic circuits.

References

[1] T. Noji, K. Kawasaki, H. Sano and N. Inoue, oral presentation, Proceeding of CARS-USA, Orlando, April 3-6(2006), pp.221-228.
[2] N. Wada, T. Hiramatsu, T. Tamura and Y. Sakabe, Abstract Book of 5th Asian Meeting on Electorceramics, Bangkok, December 10-14 (2006), p.41.
[3] Y. Higuchi, Y. Sugimoto and J. Harada, J. Eur. Ceram. Soc., 27, 2785-2788(2007).
[4] H. Niimi, K. Mihara and Y. Sakabe, J. Am. Ceram. Soc., 90, 1817-1821(2007).

EFFECT OF Mn_2O_3 ADDITION ON THE MICROSTRUCTURE AND ELECTRICAL PROPERTIES OF LEAD-FREE $Ba(Sn_{0.02}Ti_{0.98})O_3$ -$(Na_{0.5}K_{0.5})NbO_3$ CERAMICS

Chun-Huy Wang
Department of Electronic Engineering,
Nan-Jeon Institute of Technology, Tainan, Taiwan 737, R.O.C.

ABSTRACT

In this paper, the $0.02Ba(Sn_{0.02}Ti_{0.98})O_3$-$0.98(Na_{0.5}K_{0.5})NbO_3$ ceramics with the addition of 0~4.0 wt.% Mn_2O_3 has been prepared following the conventional mixed oxide process. The disc samples were sintered in alumina crucible at 1100 °C for 3 h in air. In the low Mn_2O_3 content region (≤ 1 wt.%), the decrease of dielectric constant, dielectric loss tangent and planar coupling factor together with the enhancement of mechanical quality factor correspond well to the feature of a hard doping effect on the electrical properties. For $0.02Ba(Sn_{0.02}Ti_{0.98})O_3$-$0.98(Na_{0.5}K_{0.5})NbO_3$ ceramics by doping 1.0 wt.% Mn_2O_3, the electromechanical coupling coefficients of the planar mode k_p and the thickness mode k_t reach 0.29 and 0.47, respectively, at the sintering of 1100 °C for 3 h. The significant rise of dielectric loss tangent and degradation of mechanical quality factor in the high Mn_2O_3 content region (≥ 2 wt.%) may be attributed to the appearance of distinct pores in the microstructure. With suitable Mn_2O_3 doping, a dense microstructure and good piezoelectric properties was obtained.

INTRODUCTION

At present, $Pb(Zr,Ti)O_3$ (PZT)-based piezoelectric ceramics are widely used because of their superior piezoelectric properties. However, the evaporation of harmful lead oxide during the preparation of Pb-contained ceramics contaminates the environment and the use of lead-based piezoelectric ceramics may be prohibited by the law in the future. Therefore, it is necessary and urgent to develop excellent lead-free materials replacing Pb-based piezoelectric ceramics. Lead-free piezoelectric materials such as sodium potassium niobate-based oxides, bismuth layer structure oxides and tungsten bronze-type oxides have been studied in order to replace PZT-based ceramics. Among them, $(Na_{0.5}K_{0.5})NbO_3$ (NKN) ceramics have been considered a good candidate for lead-free piezoelectric ceramics because of its strong piezoelectricity and ferroelectricity.

The hot pressed NKN ceramics (~99% of theoretical density) have been reported to possess large piezoelectric longitudinal response (d_{33}~160 pC/N), high planar coupling coefficient (k_p~45%) and high phase transition temperature (T_c = 420°C).[1] It has received a lot of attention and been thoroughly investigated.[2,3] Nevertheless, dense NKN ceramics are difficultly obtained since their phase stability is limited to 1140°C close to the melting point[4]. Moreover, they would deliquesce once exposed to humidity due to the formation of extra phases. Many researchers used hot pressing or spark plasma sintering (SPS) techniques to yield better quality ceramics[5]. Recently, an efficient solution to improve these problems has been realized by using some additives in NKN ceramics, such as ZnO,[6] $BaTiO_3$,[7] $LiNbO_3$,[8] $LiTaO_3$,[9] $SrTiO_3$,[10,11] and $CaTiO_3$.[12,13] A comparison of the properties of $(Na_{0.5}K_{0.5})NbO_3$ ceramics based on the previous reports of various groups is shown in Table I. Thus, the addition of some perovskite compounds to form solid solutions with NKN or synthesizing by Spark Plasma Sintering (SPS) has been made to obtain lead-free materials suitable for industrial applications. Recent electromechanical studies[14,15] demonstrated that $Ba(Sn_xTi_{1-x})O_3$ (BST) is basically suitable as environmental-friendly material for electromechanical sensor and actuator applications. Adding Mn is an often-adopted strategy to tailor the electrical properties of PZT-based piezoelectric ceramics.[16] Despite of these investigations, the role of Mn on the structure and electrical properties of NKN-based ceramics remains somewhat ambiguous. Therefore, further investigation on this subject is necessary.

The aim of this study is to investigate the dielectric and piezoelectric properties of

0.98(Na$_{0.5}$K$_{0.5}$)NbO$_3$-0.02Ba(Sn$_x$Ti$_{1-x}$)O$_3$ (NKN-BST) system. The effect of the dielectric and piezoelectric properties of 0.02Ba(Sn$_{0.02}$Ti$_{0.98}$)O$_3$-0.98(Na$_{0.5}$K$_{0.5}$)NbO$_3$ ceramics added with various amounts of Mn$_2$O$_3$ were discussed in this work, and the influence of the Mn$_2$O$_3$ addition on the microstructure and electrical properties was examined.

EXPERIMENTAL PROCEDURE

The 0.98(Na$_{0.5}$K$_{0.5}$)NbO$_3$-0.02Ba(Sn$_x$Ti$_{1-x}$)O$_3$ ceramics were prepared by solid-state reaction method. Additive amounts of Mn$_2$O$_3$ were added to the 0.02Ba(Sn$_{0.02}$Ti$_{0.98}$)O$_3$ -0.98(Na$_{0.5}$K$_{0.5}$)NbO$_3$ ceramics powders in concentration varying from 0 to 4.0 wt.%.

Starting materials were K$_2$CO$_3$, Na$_2$CO$_3$, BaCO$_3$, Nb$_2$O$_5$, TiO$_2$, SnO$_2$ with purities of at least 99.5% and weighted according to the stoichiometric composition (x= 0, 0.02, 0.04, 0.06 and 0.08). Additive amounts of Mn$_2$O$_3$ were added after the calcination at 900°C for 10 h. The calcined powders were pressed (CIP) at 180 MPa into pellets with 15 mm in diameter. The disc samples were sintered in alumina crucible at 1100°C for 3 h in air atmosphere. To measure relevant piezoelectric properties, the prepared ceramic samples were polarized in silicone oil at 150°C under the electric field of 4 kV/mm for 30 min. An X-ray diffractometer (Seimens D5000) using Cu Kα radiation was used to evaluate the crystal structure of the sintered ceramics. The room temperature dielectric constant was measured by LCR meter (Angilent 4284A) at 1 kHz. The piezoelectric properties were measured by a resonance-antiresonance method based on IEEE standards[17] using an impedance/gain-phase analyzer (Angilent 4194). The equation for calculating electromechanical coupling factor k_p and k_t is

$$\frac{1}{k_p^2} = 0.395\frac{f_r}{f_a - f_r} + 0.574$$

and

$$\frac{1}{k_t^2} = 0.81\frac{f_r}{f_a - f_r} + 0.405 \quad , \text{respectively} ,$$

where f_r is a resonant frequency and f_a is an anti-resonant frequency. The samples for observation of the microstructure were polished and thermally etched. The microstructures were observed using a scanning electron microscope (SEM). The mean grain size was calculated by the line intercept method[18]. The density was measured by Archimedes method.

RESULTS AND DISCUSSION

Figure 1 shows the X-ray diffraction (XRD) patterns of 0.98(Na$_{0.5}$K$_{0.5}$)NbO$_3$- 0.02Ba(Sn$_x$Ti$_{1-x}$)O$_3$ (abbreviated 0.98NKN-0.02BST) ceramics for x=0.02, 0.04, 0.06 and 0.08. All of the compositions are sintered in an air environment at 1100°C for 3 h. The orthorhombic symmetry of 0.98NKN-0.02BST ceramics at room temperature is characterized on the XRD patters in the 2θ ranges of 44-48°. The XRD analysis of sintered samples shows that 0.98NKN-0.02BST is sure of a single phase with a perovskite structure and forms a solid solution. Only an orthorhombic phase with a perovskite structure was found. The BST appears to have diffused into the NKN lattice to form a solid solution, in which Ba occupies (Na,K) lattice and (Ti,Sn) enters Nb sites of NKN. The temperature dependence of the dielectric constant at 1 kHz for 0.98NKN-0.02BST ceramics is shown in Fig. 2. For pure NKN, two sharp phase transitions are reported at 420°C and 200°C, corresponding to the phase transitions of paraelectric (cubic)–ferroelectric (tetragonal; T$_c$) and tetragonal– orthorhombic(T$_{T-O}$), respectively.[7] With increasing BST content, the paraelectric (cubic) - ferroelectric (tetragonal) and tetragonal-orthorhombic transition temperatures all shift to lower temperatures.

The physical and electrical properties of 0.98NKN-0.02BST ceramics sintered at 1100 °C for different Sn compositions were as shown in Table I. The relative density is the ratio of the measured density to the theoretical density. The relative density of 0.98NKN-0.02BST ceramic increased with increasing Sn content after sintering at 1100 °C for 3 h. For 0.98(Na$_{0.5}$K$_{0.5}$)NbO$_3$-0.02Ba(Sn$_{0.06}$Ti$_{0.94}$)O$_3$ ceramics, the electromechanical coupling coefficients of the planar mode k_p and the thickness mode k_t reach 0.18 and 0.46, respectively, after sintering at 1100 °C for 3 h. The ratio of the thickness coupling coefficient to the planar coupling coefficient is 2.56. The highest planar coupling factor (k_p = 0.31) is found for 0.98(Na$_{0.5}$K$_{0.5}$)NbO$_3$-0.02Ba(Sn$_{0.02}$Ti$_{0.98}$)O$_3$ ceramics.

Figure 3 shows the XRD patterns of 0.98(Na$_{0.5}$K$_{0.5}$)NbO$_3$-0.02Ba(Sn$_{0.02}$Ti$_{0.98}$)O$_3$ (abbreviated 0.98NKN-0.02BST2) ceramic system sintered at 1100°C for 3 h with addition of different Mn$_2$O$_3$ doping: (a) 0 wt.% (b) 0.5 wt.% (c) 1 wt.% (d) 2 wt.% (e) 4 wt.%. The specimens added with Mn$_2$O$_3$ maintain an orthorhombic phase. It indicates that the addition of small amounts of Mn$_2$O$_3$ did not give rise to an obvious change in crystal structure.

Figure 4 shows the SEM micrographs of 0.98NKN-0.02BST2 ceramic sintered at 1100°C for 3 h with various Mn$_2$O$_3$ contents: (a) 0 wt.% (b) 0.5 wt.% (c) 1 wt.% (d) 2 wt.% (e) 4 wt.%. It can be seen that the 0.98NKN-0.02BST2 ceramic consist of small grains, with a loose structure, and a high porosity in Fig. 4(a). However, the 0.98NKN-0.02BST2 ceramic with addition of 1 wt.% Mn$_2$O$_3$ doping are more dense, with low porosity, and exhibit a more homogeneous and grain size of ~2 μm in Fig. 4(c). In Fig. 4(d) and 4(e), there is a decrease in grain size with enhancing Mn$_2$O$_3$ content. Moreover, it can be observed that distinct pores appeared in the triangle regions formed by the grains in the specimens with relatively high Mn$_2$O$_3$ contents (above 2 wt.%). This implies that within the amount range of Mn$_2$O$_3$ addition in the present work, Mn mainly dissolved into the perovskite structure, because the accumulation of Mn at the grain boundaries would inhibit grain growth. These results reveal that the Mn$_2$O$_3$ addition caused a significant evolution in microstructure.

Figure 5 shows the measured density and dielectric loss tangent of 0.98NKN-0.02BST2 ceramic with various Mn$_2$O$_3$ contents sintered at 1100°C for 3 h. The measured density of the sintered samples is 90–96% of the theoretical density. The measured density increases with an increase of Mn$_2$O$_3$ contents until it reaches a maximum value at 1 wt.%, then decrease for higher Mn$_2$O$_3$ contents. The variation trend of the dielectric loss tangent (tan δ) with Mn$_2$O$_3$ content is inverse to that of the measured density. The planar coupling factor (k_p) and thickness coupling factor (k_t) of 0.98NKN-0.02BST2 ceramic with various Mn$_2$O$_3$ contents sintered at 1100°C for 3 h in Fig. 6. The electromechanical coupling factor has been used extensively as a measure of the piezoelectric response of PZT type ceramics. It was found that the electromechanical coupling factor depended on the material parameters[19] such as grain size, porosity, and chemical composition. Usually, the piezoelectric activity increases with the value of the electromechanical coupling factor in radial mode k_p and in thickness mode k_t[20]. For 0.98NKN-0.02BST2 ceramic, the k_p and the k_t reach 0.31 and 0.39, respectively, at the sintering of 1100°C for 3 h. The k_t increases with an increase of Mn$_2$O$_3$ contents until it reaches a maximum value at 1 wt.%, then decreases for higher Mn$_2$O$_3$ contents. The k_p decreases with an increase Mn$_2$O$_3$ contents.

Figure 7 shows the dielectric constant and mechanical quality factor of 0.98NKN-0.02BST2 ceramic with various Mn$_2$O$_3$ contents sintered at 1100°C for 3 h. The mechanical quality factor increases with an increase Mn$_2$O$_3$ contents until it reaches 1 wt.%, then decreases for 2 wt.%. The dielectric constant decreases with an increase Mn$_2$O$_3$ contents. The variation trend of mechanical quality factor (Q_m) with Mn$_2$O$_3$ content is inverse to that of dielectric loss tangent (tan δ), which attain a maximum value of 103 and a minimum value of 8.5% at 1 wt.% Mn$_2$O$_3$ contents, respectively. Figure 8 shows the low frequency constant in radial mode (N$_p$) and high frequency constant in thickness mode (N$_t$) of 0.98NKN-0.02BST2 ceramic with various Mn$_2$O$_3$ contents sintered at 1100°C for 3 h. The N$_p$ and N$_t$ increase with an increase of Mn$_2$O$_3$ contents until they reach a maximum value at 1 wt.%, then decrease for higher Mn$_2$O$_3$ contents.

In this research, Mn was introduced into 0.98NKN-0.02BST2 composition in the form of Mn^{3+}

Mn^{3+} have cationic radii of 0.66 Å close to that of Ti^{4+} (0.68 Å), Sn^{4+} (0.69 Å), and Nb^{5+} (0.64 Å). Thus, Mn^{3+} can enter into the octahedral site of the perovskite structure to substitute for Ti^{4+}, Sn^{4+}, and Nb^{5+} because of radius matching. Accompanying this occurrence, oxygen vacancies were created to maintain electrical neutrality. Similar to the case of PZT-based piezoelectric ceramics, the incorporation of Mn into the perovskite structure as an acceptor can generate a hard effect on the electrical properties. On the other hand, the formation of oxygen vacancies is beneficial for the mass transport during sintering. This is presumably responsible for the promoted grain growth with the Mn$_2$O$_3$ addition. The variation of the electrical properties with the addition of Mn$_2$O$_3$ can be tentatively interpreted with respect to the doping effect and microstructural evolution. When the addition amount of Mn$_2$O$_3$ is relatively low (≤1 wt.%), the hard doping effect on the electrical properties appears to be predominant. In the low Mn$_2$O$_3$ content region (≤1 wt.%), the decrease of dielectric constant, dielectric loss tangent and planar coupling factor together with the enhancement of mechanical quality factor correspond well to the feature of a hard doping effect on the electrical properties. In the case of low Mn$_2$O$_3$ content (≤1 wt.%), the grain growth became remarkable. The increase of grain size favors improving piezoelectric properties and dielectric constant, which is known as grain size effect[19]. The grain size effect compensates the decrease of planar coupling factor and dielectric constant due to the hard doping effect. This is assumedly responsible for the slight fall of planar coupling factor and dielectric constant in the low Mn$_2$O$_3$ content region (≤1 wt.%). The significant rise of dielectric loss tangent and degradation of mechanical quality factor in the high Mn$_2$O$_3$ content region (≥2 wt.%) may be attributed to the appearance of distinct pores in the microstructure. Therefore, it is likely that the doping and microstructural effects contribute to the electrical properties of the ceramics in a cooperative way. With increasing the porosity of 0.98NKN-0.02BST2 ceramic, the piezoelectric and dielectric properties will diminish gradually.

CONCLUSION

The 0.98NKN-0.02BST solid solution ceramics were prepared by the conventional ceramics technique. From the XRD patterns, the orthorhombic symmetry of 0.98NKN-0.02BST ceramic at room temperature was found.

The microstructure and electrical properties of 0.98NKN-0.02BST2 ceramic with the addition of 0~4.0 wt.% Mn$_2$O$_3$ were discussed. In the low Mn$_2$O$_3$ content region (≤1 wt.%), the decrease of dielectric constant, dielectric loss tangent and planar coupling factor together with the enhancement of mechanical quality factor correspond well to the feature of a hard doping effect on the electrical properties. For 0.98NKN-0.02BST2 ceramic by doping 1.0 wt.% Mn$_2$O$_3$, the electromechanical coupling coefficients of the planar mode k_p and the thickness mode k_t reach 0.29 and 0.47, respectively, at the sintering of 1100°C for 3 h. When the content of Mn$_2$O$_3$ was above the solubility limit of 2.0 wt.%, Mn ions accumulated at grain boundary and inhibited the grain growth, which increased the volume of space charge regions and deteriorated the piezoelectric properties. The significant rise of dielectric loss tangent and degradation of mechanical quality factor in the high Mn$_2$O$_3$ content region (≥2 wt.%) may be attributed to the appearance of distinct pores in the microstructure. With suitable Mn$_2$O$_3$ doping, a dense microstructure and good piezoelectric properties was obtained.

ACKNOWLEDGEMENT

The author would like to acknowledge the financial support of the National Science Council (under contract NSC-96-2215-E-232-002) of the Republic of China.

REFERENCES
[1]G. Shirane, R. Newnham, and R. Pepinsky, Dielectric properties and phase transitions of NaNbO$_3$ and

(Na, K)NbO$_3$, Phys. Rev., **96**, 581–588 (1954).

[2]V. Lingwal, B. S. Semwal, N.S. Panwar, Dielectric properties of Na$_{1-x}$K$_x$NbO$_3$ in orthorhombic phase, Bull. Mater. Sci., **26**, 619–625 (2003).

[3]H. Birol, D. Damjanovic, and N. Setter, Preparation and characterization of (K$_{0.5}$Na$_{0.5}$)NbO$_3$ ceramics, J. Eur. Ceram. Soc., **26**, 861–866 (2006).

[4]S. Y. Chu, W. Water, Y.D. Juang, J. T. Liaw, and S. B. Dai, Piezoelectric and dielectric characteristics of lithium potassium niobate ceramic system, Ferroelectrics, **297**, 11–17 (2003).

[5]M. Matsubara, T. Yamaguchi, K. Kikuta, and S. Hirano, Effect of Li substitution on the piezoelectric properties of potassium sodium niobate ceramics, Jpn. J. Appl. Phys., **44** 6136–6142 (2005).

[6]S. H. Park, C. W. Ahn, S. Nahm, and J. S. Song, Microstructure and piezoelectric properties of ZnO-added (Na$_{0.5}$K$_{0.5}$)NbO$_3$ ceramics, Jpn. J. Appl. Phys. **43**, L1072–L1074 (2004).

[7]Y. Guo, K. Kakimoto, and H. Ohsato, Structure and electrical properties of lead-free (Na$_{0.5}$K$_{0.5}$)NbO$_3$–BaTiO$_3$ ceramics, Jpn. J. Appl. Phys., **43**, 6662 (2004).

[8]Y. Guo, K. Kakimoto, and H. Ohsato, Phase transitional behavior and piezoelectric properties of (Na$_{0.5}$K$_{0.5}$)NbO$_3$–LiNbO$_3$ ceramics, Appl. Phys. Lett., **85**, 4121–4123 (2004).

[9]Y. Guo, K. Kakimoto, H. Ohsato, (Na$_{0.5}$K$_{0.5}$)NbO$_3$–LiTaO$_3$ lead-free piezoelectric ceramics, Mater. Lett., **59**, 241–244 (2005).

[10]M. Kosec, V. Bobnar, M. Hrovat, J. Bernard, B. Malic, and J. Holc, New lead-free relaxor based on the K$_{0.5}$Na$_{0.5}$NbO$_3$–SrTiO$_3$ solid solution, J. Mater. Res., **19**, 1849–1854 (2004).

[11]Y. Guo, K. Kakimoto, and H. Ohsato, Dielectric and piezoelectric properties of lead-free (Na$_{0.5}$K$_{0.5}$)NbO$_3$–SrTiO$_3$ ceramics, Solid State Commun., **129**, 279–284 (2004).

[12] R. C. Chang, S. Y. Chu, Y. F. Lin, C. S. Hong, and Y. P. Wong, An investigation of (Na$_{0.5}$K$_{0.5}$)NbO$_3$–CaTiO$_3$ based lead-free ceramics and surface acoustic wave devices, J. Eur. Ceram. Soc. **27**, 4453-4460, (2007).

[13] R. C. Chang, S. Y. Chu, Y. F. Lin, C. S. Hong, P. C. Kao, and C. H. Lu, The effects of sintering temperature on the properties of (Na$_{0.5}$K$_{0.5}$)NbO$_3$–CaTiO$_3$ based lead-free ceramics, Sens. Actuators A, **138**, 355-360 (2007).

[14]V. Cieminski, J. Langhammer, H. T. Abicht, and H. P. Peculiar, Electromechanical properties of some Ba(Ti,Sn)O$_3$ ceramics, Phys. Stat. Sol (a), 120, 285–293 (1990).

[15] V. Mueller, A. Kouvatov, R. Steinhausen, H. Beige, and H. P. Abicht, Ferroelectric and relaxor-like electromechanical strain in BaTi$_{1-x}$Sn$_x$O$_3$ ceramics, Integrat. Ferroelectrics, 63, 593–596 (2004).

[16]E. Boucher, B. Guiffard, L. Lebrun, D. Guyomar, Effects of Zr/Ti ratio on structural, dielectric and piezoelectric properties of Mn- and (Mn, F)-doped lead zirconate titanate ceramics, Ceramics International, **32**, 479–485 (2006).

[17]Anon., IRE Standards on Piezoelectric Crystals: Measurement of Piezoelectric Ceramics, Proc. IRE., **49**, 1161-1168(1961).

[18]T. Senda, and R. C. Bradt, Grain Growth in Sintered ZnO and ZnO-Bi$_2$O$_3$ Ceramics, J. Am. Ceram. Soc., **73** [1] 106-114 (1990).

[19]K. Okazaki and K. Nagata, Effects of Grain Size ad Porosity on Electrical and Optical Properties of PLZT Ceramics, J. Am. Ceram. Soc. **56** [2] 82-86 (1973).

[20]K. H. Hartal, Physics of Ferroelectric Ceramics used in Electronic Devices, Ferroelectrics, **12**, 9-19 (1976)

[21]V. A. Isupov, and Yu. E. Stolypin, Coexistence of Phases in Lead Zirconate-Titanate Solid Solutions, Sov. Phys.-Solid State, **12**, 2067-71 (1970).

[22] E. G. Fesenko, A. Y. Dantsiger, L. A. Resnitohenko, and M. F. Kupriyanov, Composition–structure-properties dependences in solid solutions on the basis of lead–zirconate–titanate and sodium niobate, Ferroelectrics, **41**, 137–141 (1982).

[23] V.A.Bokov, Dielectric Properties of PbTiO$_3$-PbZrO$_3$ Ceramics, Sov.Phys- Tech.Phys., **2**, 1657-62

(1957).

Table I. Comparison of properties of (Na$_{0.5}$K$_{0.5}$)NbO$_3$ ceramics based on previous reports of various groups.

Composition	Measured density (g/cm^3)	Relative density (%)	K$_o$	k_p	k_t	k_t /k_p	Ref.
NKN	4.34	96.4	---	0.295	0.41	1.39	7
NKN-ZnO	4.26	94.5	500	0.4	---	---	6
0.98NKN-0.02BaTiO$_3$	4.44	98.4	---	0.29	0.38	1.31	7
0.995NKN-0.005SrTiO$_3$	4.44	98.4	412	0.325	0.438	1.35	10
0.94NKN-0.06LiNbO$_3$	4.35	96.5	---	0.42	0.48	1.14	8
0.94NKN-0.06LiTaO$_3$	---	---	570	0.36	---	---	9
0.995NKN-0.005CaTiO$_3$	4.4	97.6	553	0.42	0.38	1.1	12
0.98NKN-0.02 Ba(Sn$_{0.02}$Ti$_{0.98}$)O$_3$	4.15	92.2	365	0.31	0.39	1.44	This study
0.98NKN-0.02 Ba(Sn$_{0.04}$Ti$_{0.96}$)O$_3$	4.20	93.3	375	0.27	0.42	1.55	This study
0.98NKN-0.02 Ba(Sn$_{0.06}$Ti$_{0.94}$)O$_3$	4.25	94.4	380	0.18	0.46	2.56	This study
0.98NKN-0.02 Ba(Sn$_{0.08}$Ti$_{0.92}$)O$_3$	4.28	95.1	410	0.19	0.41	2.18	This study

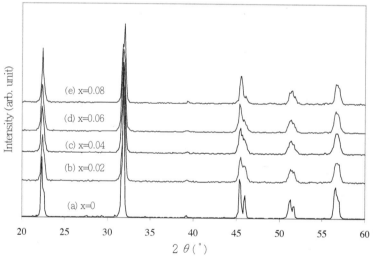

Figure 1.   XRD patterns of 0.98(Na$_{0.5}$K$_{0.5}$)NbO$_3$-0.02Ba(Sn$_x$Ti$_{1-x}$)O$_3$ ceramic system for differenr x compositions: (a) x=0, (b) x=0.02, (c) x=0.04, (d) x=0.06, and (e) x=0.08.

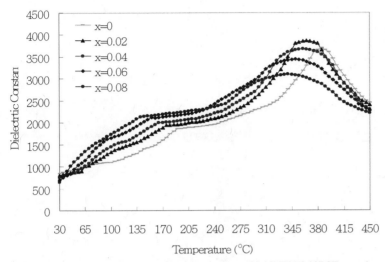

Figure 2. Temperature dependence of dielectric constant of 0.98NKN-0.02BST ceramic at 1 kHz.

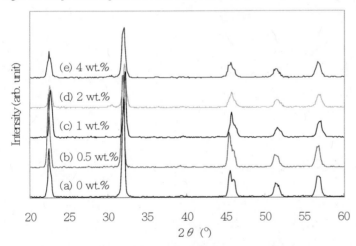

Figure 3. XRD Patterns of 0.98NKN-0.02BST2 ceramic after sintering at 1100°C for 3 h with addition of different Mn₂O₃ doping: (a) 0 wt.% (b) 0.5 wt.% (c) 1 wt.% (d) 2 wt.% (e) 4 wt.%.

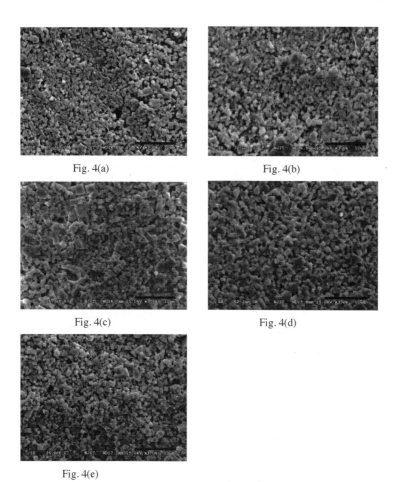

Fig. 4(a)   Fig. 4(b)

Fig. 4(c)   Fig. 4(d)

Fig. 4(e)

Figure 4. SEM images of 0.98NKN-0.02BST2 ceramic with addition of different Mn$_2$O$_3$ doping after sintering at 1100°C for 3 h: (a) 0 wt% (b) 0.5 wt% (c) 1 wt% (d) 2 wt% (e) 4 wt%. Bar=10 μm

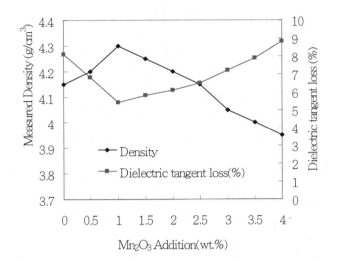

Figure 5. Density and dielectrid tangent loss of 0.98NKN-0.02BST2 ceramic with addition of different Mn₂O₃ doping after sintering at 1100°C for 3 h.

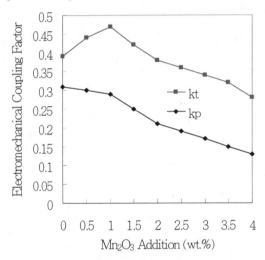

Figure 6. Electromechanical coupling factor of 0.98NKN-0.02BST2 ceramic with addition of different Mn₂O₃ doping after sintering at 1100°C for 3 h.

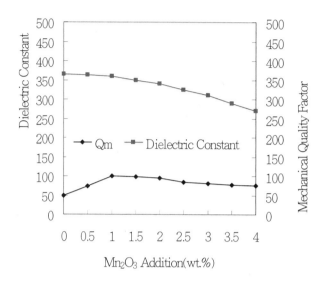

Figure 7. Dielectric constant and mechanical quality factor of 0.98NKN-0.02BST2 ceramic with addition of different Mn$_2$O$_3$ doping after sintering at 1100°C for 3 h.

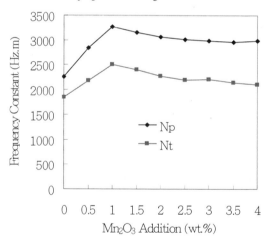

Figure 8.   Frequency constant of 0.98NKN-0.02BST2 ceramic with addition of different Mn$_2$O$_3$ doping after sintering at 1100°C for 3 h.

ELECTRONIC PROPERTIES OF BaTiO$_3$ CONTAINING GLASS CERAMICS

M. Letz[a], M. Aigner[a], M.J. Davis[b], B. Hoppe[a], M. Kluge[a],
I. Mitra[a], B. Rüdinger[a], D. Seiler[a]

[a]Schott AG, Research and Development, D-55014 Mainz, Germany
[b]SCHOTT R&D North America, Dureya, USA

ABSTRACT

Glass ceramics are a class of material which are produced via a glass melting route. In a second step crystalline grains, which are closely and ideally pore free embedded in a residual glass matrix, are grown in a ceramization process. A glass ceramic with nanocrystalline grains of BaTiO$_3$ as a ferroelectric phase shows a large dielectric constant together with – opposite to ceramic BaTiO$_3$ – a flat temperature characteristic of its dielectric properties. Together with the expected large breakdown strength in a pore free material outstanding dielectric properties are obtained. We show results of BaTiO$_3$ containing glass ceramics and discuss their electronic properties. We argue that the size of the ferroelectric domains being in the order of a few nm allows us to reach the large dielectric constants indicating superparaelectricity.

INTRODUCTION

Under a glass ceramic we understand a material which is obtained via a homogeneous melt. The melt is supercooled into a glassy state, which allows for a number of hot forming processes. In a second and independent step the glass is treated with a well defined time-temperature profile to allow controlled growth of crystals within the residual glass. The most common example for such glass ceramic are Li-Al-Si glass ceramics [Stookey1959] LAS glass ceramics, which are well known as brand names like Ceran [Ceran], Robax [Robax] or Zerodur [Zerodur]. A further example for such glass ceramics are Y-Al-Si glass ceramics. They contain YAG (Y$_3$Al$_5$O$_{12}$) as the only crystalline phase. Doped with Ce they can be used as a blue to yellow converter in white LED's [Fujita2008, Engel2007]. Also perovskites esspecially BaTiO$_3$ are well known as crystalline phases in glass ceramics [Herczog1964]. Despite their long history they never reached commerciallization as a mass product. BaTiO$_3$ ceramics are however applied in large scale in ceramic capacitors. This is due to the existence of ferroelectricity in BaTiO$_3$. Codopands are well known to alter for ceramic BaTiO$_3$ the temperature dependence of the dielectric constant as well as to minimize the dielectric loss [Jaffe1971]. Also the role of grain size in these ceramics is deeply investigated [McNeal1998] even though the grain growth during sintering does not really allow to obtain nanometer sized crystallites.

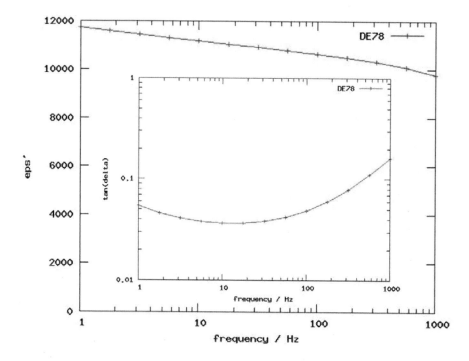

Fig.1: The frequency dependent dielectric constant of a glass ceramic with nano crystal grains is plotted. Dielectric constants of the order of e'=10000 are observed. The decay which starts around 1 kHz we assign to the boundary frequency of our measurement conditions since the impedance gets very small at higher frequencies leading to very large currents. The inset shows the dielectric loss tangent, $\tan\delta=\varepsilon''/\varepsilon'$.

Turning back to glass ceramics, the main drivers to obtain a particular crystalline phase with a defined size of crystallites and a defined size distribution are (i) the thermodynamic equilibrium of underlying crystalline phases, (ii) the kinetic of crystal growth and (iii) the nucleation mechanism and number density of nucleating seeds. The main task to obtain a ferroelectric glass ceramic in the Ba-Ti-Si system is to avoid the ubiquous non ferroelectric Fresnoite ($BaTi_2Si_2O_8$) as a crystal phase. According to thermodynamic equilibrium this phase can always occur if one adds SiO_2 into a melt of $BaTiO_3$ [Rase1955]. Only crystallization kinetic allows, as shown in the present work, to obtain glass ceramics with $BaTiO_3$ as the only crystalline phase. We further demonstrate that a control of ceramization parameters allows to obtain crystallites whose size is of the order of the ferroelectric domaines. In this way glass ceramics are able to enter the superparaelectric regime, where the crystallites have diameters which are of the order of the size of the ferroelectric domaines and where the orientation of the large polarization of a single grain can be altered by only thermal fluctuations.

SINGLE CRYSTAL BaTiO₃

Ceramic BaTiO₃ shows, if the size of the crystallites is much larger than the size of the ferroelectric domaines, mostly the dieletric material properties of single crystalline material. There are several structural phase transition at normal pressure as a function of temperature. There is a high temperature cubic phase which undergoes a weak first order transition at 130 °C (303 K) from a cubic to tetragonal symmetry. Around 0 °C (273 K) there is a second ferroelectric transition when the crystal symmetry changes from tetragonal to orthorhombic [Jaffe1971]. Around -90 °C (183 K) an orthorhombic to rhombohedral phase transition can be found. The enormous importance of BaTiO₃ is due to the fact that room temperature is always embedded within the two ferroelectric transitions at 0 °C and at 130 °C. The ferroelectric transitions are accompanied by large electric polarizabilities which leads to dielectric constants ε'>1000. Codoping e.g. with Sr or La changes the dielectric properties by flattening the temperature chracteristic or by reducing the dielectric loss. In the ferroelectric phase, ferroelectric domaines are formed with domain diameters of approx. 30-50nm and a predominatly 90° shift of the polarization direction of neighboring domaines. The domain wall movement together with a clamping of the 90° domain walls, which leads to the emission of GHz shear waves [Arlt1993], are the dominant mechanism for dielectric loss at high frequencies up to the GHz spectral range

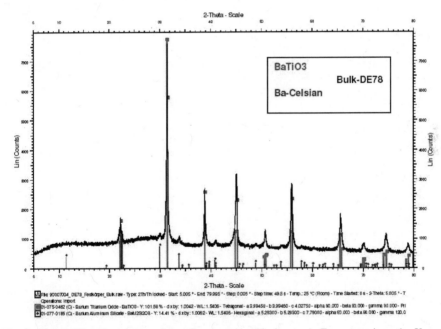

Fig. 2: X-ray diffraction pattern of the glass ceramic (black curve). For comparison the X-rax diffraction patterns of ceramic BaTiO₃ and B-Celsian are shown. The main phase is BaTiO₃ with very small traces of Ba-Celsian.

GLASS CERAMIC

When making glass ceramics with $BaTiO_3$ as the crystalline phase in the Ba-Ti-Si system, the following problems occur. If the composition is too close to stoechiometric $BaTiO_3$, the melting temperature raises to very high temperatures and the glass stability is poor. When adding SiO_2 to the melt the glass stability increases but Fresnoite ($Ba_2TiSi_2O_8$) occurs as a second crystal phase. Adding Al_2O_3 as a second glass network forming oxide can give Ba-Celsian ($Ba_2AlSi_2O_8$) as a further crystal phase. Inbetween these boundary conditions is the parameter regime for obtaining stable glasses, which can be ceramized under further heat treatment. In previous studies on these systems there where either a small phase content of ferroelectric $BaTiO_3$ [Herczog1964] or the $BaTiO_3$ was accompanied by other non ferroelectric crystal phases like Fresnoite or Ba-Celsian [McCauley1998]. Therefore the dielectric constant of these glass ceramics never exceeded values of $\varepsilon' \sim 1000$.

EXPERIMENT

BaO and TiO_2 was melted together with glass formers (SiO_2 and Al_2O_3). The melt was cooled down in order to obtain a homogeneous glass. The glass was than heat treated in a controlled way in order to obtain a particular crystalline microstructure. The ramp up speed of the temperature, the temperatures and the holding times are crucial parameters to obtain a particular grain size distribution. X-ray diffraction on a powdered part of the sample was made and a part was prepared for dielectric measurements by contacting with silver paste, which was typically 20mm in diameter and 1mm thick.

RESULTS AND DISCUSSION: GLASS CERAMIC WITH NANO-CRYSTALLINE GRAINS

Glass ceramics with $BaTiO_3$ as the only crystal phase in high volume content with nano crystals have a number of advantages. First each individual crystallite forms a single ferroelectric domain which can be easily, by only thermal fluctuations, oriented in a external electric field. This leads to very high dielectric constants and is called [McNeall1998] the superparaelectric effect. The superparaelectric effect is described in theory [Bell1993] and is discussed for a number of ferroelectric perovskite materials like $PbTiO_3$ [Ruediger2005, Roelofs2003] or La doped PbZrTiO [Viehland1992].

In Fig. 1 we plot the frequency dependent dielectric constant of a glass ceramic with nano crystals of $BaTiO_3$. For orienting the polarization direction of such a small crystal no domain walls have to be shifted. The X-ray diffraction pattern is shown in Fig. 2 which clearly shows that $BaTiO_3$ is the only crystal phase with the exception of small traces of Ba-celsian. When measuring the dielectric constant we are aware of surface polarization mechanism [Lunkenheimer2002] which we tried to exclude by different surface treatments.

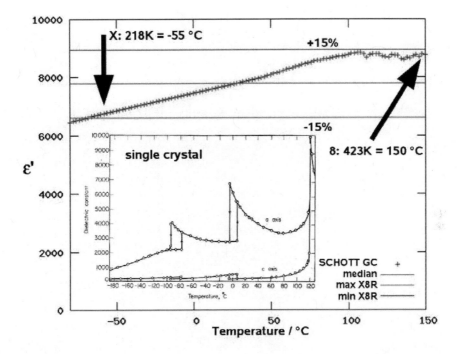

Fig.3 The temperature dependence of the dielectric constant is plotted for a frequency of 1 kHz. Within the temperature range of -55 °C and 150 °C the dielectric constant stays within the +/-15 % range. A capacitor build from such a glass ceramic will fulfil X8R standard for electronic industry. In the inset we show the temperature dependence of the orientation dependent dielectric constant for single crystal BaTiO₃ from [Jaffe 1971].

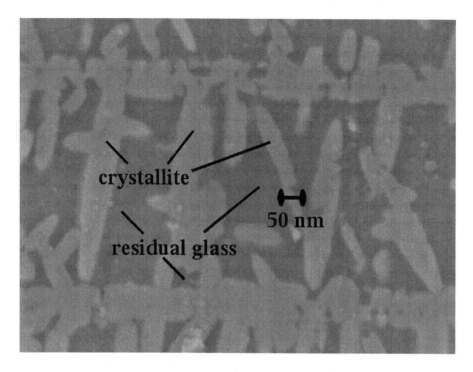

Fig.4 The microstructure is shown. The crystallites are mostly needles which have a length of up to 200nm and a diameter of approximately 50nm. The amount of crystalline phase is in the range of 50 - 70 volume %

MICROSTRUCTURE – NANOSTRUCTRE

Also the temperature dependence of the dielectric constant is expected to be fairly flat, since the existence of the small crystallites, which have diameters in the order of the ferroelectric domains, smears out the sharp ferroelectric transitions. In Fig. 3 we plot the temperature dependence of the dielectric constant of one of our glass ceramic sample with the frequency as a parameter. We stress that the room temperature dielectric measurement in Fig. 1 was made on a different sample, with different number density of the grains and with different grain size distribution than the temperature dependent measurement shown in Fig. 3. Therefore the absolute number of the dielectric constant also differs between the two samples.

For the frequency of 1 kHz we include in the plot horizontal lines which mark +/-15 % of the average value within the temperature interval from -55 °C to 150 °C. Within this range a capacitor made from our glass ceramic will fulfil the X8R standard for ceramic capacitors in electronic industry.

Also the temperature dependence of the dielectric constant is expected to be fairly flat, since the existence of the small crystallites, which have diameters in the order of the ferroelectric domains, smears out the sharp ferroelectric transitions. In Fig. 3 we plot the temperature dependence of the dielectric constant of one of our glass ceramic sample with the frequency as a parameter. We stress that the room temperature dielectric measurement in Fig. 1 was made on a different sample, with different number desity of the grains and with different grain size distribution than the temperature dependent measurement shown in Fig. 3. Therefore the absolute number of the dielectric constant also differs between the two samples.

For the frequency of 1 kHz we include in the plot horizontal lines which mark +/-15 % of the average value within the temperature interval from -55 °C to 150 °C. Within this range a capacitor made from our glass ceramic will fulfill the X8R standard for ceramic capacitors in electronic industry.

In Fig 4. we show, using an electron microscopic image, that the structure of our glass ceramic contains small crystallites, many of them are needle shaped with lengths of the order of 100nm and diameter of the order of 20-30nm. Such a microstructure is competely different from the microstructure found in ceramic materials. According to our understanding this microstructure – or better nanostructure – is the essential key to understand the outstanding dielectric properties of our material. The volume fraction of the crystal phase lies in the order of 50% - 80%.

CONCLUSION

Glass ceramics with BaTiO₃ as a crystal phase allow to adjust the size of the crystalline grains in the order of the size of ferroelectric domains. This means the size of the crystalline grains is of the order of ~50 nm. In this way the superparaelectric regime can be reached where a reorientation of the polarizability of a single grain is possible by thermal fluctuations only. This leads to a very large polarizability of $\varepsilon' \sim 10000$ which shows only a weak temperature dependence. The flat temperature dependence suggests that it is possible to reach the X8R standard of electronic industry with capacitors build from these glass ceramics. Obtained from a homogeneous glassy phase, the glass ceramics are pore free and expect to show high dielectric breakdown strengths.

ACKNOWLEDGEMENT

M.L. acknowledges discussions with P. Lunkenheimer from Augsburg University and G. Schaumburg and D. Wilmer from the company Novocontrol on dielectric measuremets.

Ps.: We have recent evidence that the dielectric data shown in Fig 1 are forged by a large contact resistance, a subject which we currently investigate.

REFERENCES

[Stookey1959] S.D. Stookey, Ind. Eng. Chem., **51**, 805 (1959).
[Ceran] Ceran®, http://www.schott.com/hometech/german/products/ceran/
[Robax] Robax®, http://www.schott.com/hometech/german/products/robax/
[Zerodur] Zerodur®,
   www.schott.com/optics_devices/german/download/zerodur_katalog_deutsch_2004.pdf
[Fujita2008] S. Fujita, A. Sakamoto, S. Tanabe, IEEE J. of sel. top. in quant. electronics, **14**, 1387
   (2008).
[Engel2007] A. Engel, M. Letz,T. Zachau, E. Pawlowski, K. Seneschal-Merz, T. Korb, D. Enseling,
   B. Hoppe, U. Peuchert, J.S. Hayden, Proc. SPIE, **6486**, Y4860 (2007).
[Herczog1964] A. Herczog, J. Am. Ceram. Soc., **47**, 107 (1964).
[Jaffe1971] B. Jaffe, W.R. Cook, H. Jaffe, *Piezoelectric ceramics* (Academic press, London, pp. 53,
   1971).
[McNeal1998] M.P. McNeal, S.-J. Jang, R.E. Newnham, J. Appl. Phys., **83**, 3288 (1998).
[Rase1955] D.E. Rase, R. Roy, J. Am. Ceram. Soc., **38**, 393 (1955).
[Arlt1993] G. Arlt, U. Böttger, S. Witte, J. Am. Ceram. Soc., **63**, 602 (1993).
[McCauley1998] D. McCauley, R.E. Newnham, C.A. Randall, J. Am. Ceram. Soc., **81**, 979 (1998).
[Bell1993] A.J. Bell, J. Phys. Cond. Mat., **5**, 8773 (1993).
[Ruediger2005] A. Rüdiger, T. Schneller, A. Roelofs, S. Tiedke, T. Schmitz, R. Waser,
   Appl. Phys. A, **80**, 1247 (2005).
[Roelofs2003] A. Roelofs, T. Schneller, K. Szot, R. Waser, Nanotechnology, **14**, 250 (2003).
[Viehland1992] D. Viehland, J.F. Li, S.J. Jang, L.E. Cross, M. Wuttig, Ohys. Rev. B, **46**, 8013
   (1992).
[Lunkenheimer2002] P. Lunkenheimer, V. Bobnar, A.V. Pronin, A.I. Ritus, A.A. Volkov,
   A. Loidl, Phys. Rev. B, **66**, 052105 (2002).

DEVELOPMENT OF (100) THREE-AXIS-ORIENTED SINGLE CRYSTAL (Ba$_{0.7}$Sr$_{0.3}$)TiO$_3$ THIN FILM FABRICATION ON Pt/MgO(100) SUBSTRATE BY CHEMICAL SOLUTION DEPOSITION METHOD

Tadasu Hosokura, Keisuke Kageyama, Hiroshi Takagi and Yukio Sakabe

Murata Manufacturing Co., Ltd.
10-1, Higashikotari, 1-chome, Nagaokakyo-shi, Kyoto 617-8555 Japan

ABSTRACT
Hetero-epitaxially grown perovskite (100) three-axis-oriented (Ba$_{0.7}$Sr$_{0.3}$)TiO$_3$ thin films were prepared on (100) platinum coated (100) magnesium oxide (MgO) single crystal substrate by the chemical solution deposition method using a solution derived from Ba(CH$_3$COO)$_2$, Sr(CH$_3$COO)$_2$, and Ti(O-i-C$_3$H$_7$)$_4$.
The growth of the film was found to depend on the heat treatment condition. A (Ba,Sr)TiO$_3$ thin film heat treated at 800 °C was found to be three-axis-oriented by TEM. The film exhibited a (100) three-axis-orientation that followed the (100) orientation of the Pt substrate, as observed from an X-ray pole figure measurement and SAED patterns. The C-V measurement and the temperature dependence of the dielectric constant were performed to reveal the electrical properties of the films.

I. INTRODUCTION
Barium strontium titanate (BST) thin films have been increasingly desired as ferroelectric materials for fabricating ferroelectric memory devices, multilayer capacitor, optical modulator, etc.[1] Recently, high quality ferroelectric thin films have been used for fabricating advanced microwave signal processing devices. Small, compact, low power microwave devices that can be fabricated from structures based on ferroelectric films include phase shifters, tunable filters, tunable resonators, phased array antennas, and frequency-agile microwave radio transreceivers. Thin films fabricated using BST meet the majority of the device requirements due to their unique combination of properties.
BST thin films have been fabricated by various techniques such as rf-sputtering[2], pulsed laser deposition[3,4], metal-organic chemical vapor deposition (MOCVD)[5], and chemical solution deposition (CSD)[6,7].
Among these methods, the CSD method has an edge over the other deposition techniques in terms of achieving good homogeneity, chemical composition control, high purity, low processing temperature, and applicability to large substrate areas, using simple and inexpensive equipment[8].
Many researchers have fabricated one-axis-oriented (Ba$_{0.7}$Sr$_{0.3}$)TiO$_3$ thin films comprising columnar grains by the CSD method[9].
We have reported that quasi hetero-epitaxially grown (Ba$_{0.7}$Sr$_{0.3}$)TiO$_3$ thin films were fabricated on a Pt-coated silicon substrate by the sol-gel method using a (Ba$_{0.7}$Sr$_{0.3}$)TiO$_3$ sol derived from Ba(CH$_3$COO)$_2$, Sr(CH$_3$COO)$_2$, and Ti(O-i-C$_3$H$_7$)$_4$[10].
Takeshima et al. have reported that (Ba$_{0.6}$Sr$_{0.4}$)TiO$_3$ thin films were fabricated by the MOCVD method and the hetero-epitaxially grown BST was deposited on a Pt(100)/MgO(100) substrate[11].
The application of BST to devices requires high quality BST films for minimizing loss tangents. The fabricated three-axis-oriented BST film can minimize the loss tangent. As the CSD method is low temperature processing, it is kinetically limited method as compared to physical vapor deposition (PVD) methods. Little research has been conducted on the fabrication of the three-axis-oriented BST films by the CSD method.
In this paper, we report the preparation of a (100) three-axis-oriented (Ba$_{0.7}$Sr$_{0.3}$)TiO$_3$ thin film on a Pt(100)/MgO(100) substrate by the CSD method and the characterization of the micro structure, crystallinity and the electrical properties of the film.
Fabricating the three-axis-oriented BST film by CSD method enables to produce dense ferroelectric thin films and high-quality electric devices.

II. EXPERIMENTAL PROCEDURE
The Pt(100)/MgO(100) substrate was used for fabricating the (Ba$_{0.7}$Sr$_{0.3}$)TiO$_3$ thin film by the CSD method. MgO (100) wafer was cleaned in alkaline detergent by a supersonic jet source and then dried by using a N$_2$ blower. The Pt film was hetero-epitaxially grown on the MgO (100) wafer at

650 °C by dc sputtering. A chemical solution was prepared by dissolving $Ba(CH_3COO)_2$, $Sr(CH_3COO)_2$, and $Ti(O\text{-}i\text{-}C_3H_7)_4$ with a molar ratio of 70 : 30 : 100 in a mixed solvent of acetic acid and ethyleneglycol monoethylether. The concentration of the solution was 0.3 M.

The chemical solution was deposited by spin coating on the $Pt(100)/MgO(100)$ substrate. This solution was dispersed on the substrate and then spin coated at 4000 rpm for 30 s. After spin coating, the films were dried on a hot plate at 423 K for 3 min. The dried and coated films were heat treated at 650 °C or 800 °C for 20 min at a heating rate of 300 °C /min by rapid thermal processing (RTP). In order to prepare thicker $(Ba_{0.7}Sr_{0.3})TiO_3$ thin films, the deposition procedure was repeated 8 times. Pt dots with a diameter of 0.5 mm were sputtered by placing a mask on the film to form metal–dielectric–metal (MDM) capacitors.

The cross-sectional images of the obtained films were taken with a field emission-scanning electron microscope (FE-SEM) (Model S-5000, Hitachi, Tokyo, Japan). Cross section images of the obtained films and the selected area electron diffraction (SAED) patterns were obtained by using a transmission electron microscope (TEM) (Model EM-002B, TOPCON, Tokyo, Japan). Crystal characterizations were performed by X-ray diffraction (XRD) (Model RINT-KI, Rigaku, Tokyo, Japan). The orientation of the film along the axis parallel to that of the substrate was characterized by an X-ray pole figure measurement (Model x'pert, Philips, Eindhoven, The Netherlands). Small signal ac (100 mV, 1 kHz) capacitance, loss tangent, and capacitance–voltage (C-V) were measured using an LCR meter (HP 4284A, Hewlett-Packard, California, USA). The temperature dependence of the dielectric constant was measured using an HP 4284A LCR meter (Model MPC, MMR Technologies, California, USA).

III. RESULTS AND DISCUSSION

Figure 1(a) shows a cross-sectional SEM image of the $(Ba_{0.7}Sr_{0.3})TiO_3$ thin film heat treated at 650 °C. The $(Ba_{0.7}Sr_{0.3})TiO_3$ thin film heat treated at 650 °C comprised 30 nm average-sized grains. The grains were randomly nucleated and crystallized during heat treatment. The thickness of the $(Ba_{0.7}Sr_{0.3})TiO_3$ thin film was confirmed to be 280 nm. Figure 1(b) shows the cross-sectional SEM image of the $(Ba_{0.7}Sr_{0.3})TiO_3$ thin film heat treated at 800 °C. This $(Ba_{0.7}Sr_{0.3})TiO_3$ thin film indicated no evidence of grain boundaries. The thickness of the $(Ba_{0.7}Sr_{0.3})TiO_3$ thin film was confirmed to be 230 nm.

Heat treatment at 800 °C gives the high ion mobilities. This enabled the formation of a high density layer and giant grain growth. In comparison, heat treated at 650 °C gives the low ion mobilities. This led to the grains being randomly nucleated, and the grain growth ceased at 30 nm. These SEM images indicated that the morphology of the films depended on the heat treating conditions as described in ref. 9.

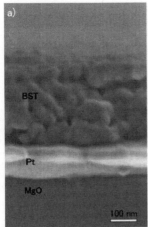 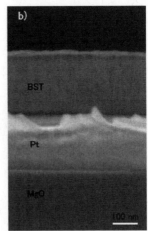

Fig. 1. (a) Cross-sectional SEM image of the BST thin film heat treated at 650 °C. (b) Cross-sectional SEM image of the BST thin film heat treated at 800 °C.

Figure 2(a) shows an XRD pattern of the $(Ba_{0.7}Sr_{0.3})TiO_3$ thin film heat treated at 650 °C. In the XRD pattern of this $(Ba_{0.7}Sr_{0.3})TiO_3$ thin film, (100) reflection and (110) reflection were observed. This result indicated that the film heat treated at 650 °C was a poly crystal perovskite $(Ba_{0.7}Sr_{0.3})TiO_3$ thin film. Figure 2(b) shows XRD pattern of the $(Ba_{0.7}Sr_{0.3})TiO_3$ thin film heat treated at 800 °C. Strong (h00) reflections were observed for this $(Ba_{0.7}Sr_{0.3})TiO_3$ thin film deposited. This result indicated that the perovskite $(Ba_{0.7}Sr_{0.3})TiO_3$ (100) thin film was hetero-epitaxially grown on the Pt(100)/MgO(100) substrate. These results indicated that the crystallinity of thin films can be controlled by varying the heat treatment conditions. The film heat treated at 650 °C was further studied by X-ray pole figure measurement and TEM to confirm its epitaxial nature.

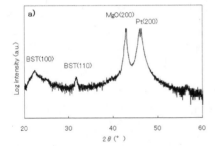 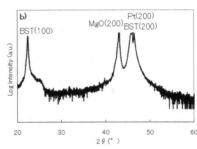

Fig. 2. (a) XRD pattern of the BST thin film heat treated at 650 °C. (b) XRD pattern of the BST thin film heat treated at 800 °C.

Figure 3 shows X-ray pole figure measured at a fixed 2-theta angle corresponding to the 110 reflection of the $(Ba_{0.7}Sr_{0.3})TiO_3$ thin film heat treated at 800 °C. Four-fold symmetry and a pole was observed at angle theta of about 45°. This result indicated that the (100) three-axis-oriented $(Ba_{0.7}Sr_{0.3})TiO_3$ thin film was deposited on the Pt(100)/MgO(100) substrate. This result agreed with the result of XRD pattern.

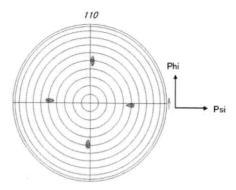

Fig. 3. X-ray pole figure measured at fixed 2-theta angle corresponding to BST 110 of the BST thin film heat treated at 800 °C.

Figure 4(a) show cross-sectional TEM images of the $(Ba_{0.7}Sr_{0.3})TiO_3$ thin film deposited on Pt(100)/MgO(100) substrate. The SAED pattern of the Pt-sputtered film, the boundary of the Pt sputtered film and the $(Ba_{0.7}Sr_{0.3})TiO_3$ thin film, and the $(Ba_{0.7}Sr_{0.3})TiO_3$ thin film are also shown in figures 4(b), 4(c), 4(d), respectively. The SAED patterns were obtained when the direction of the electron beam was parallel to the [010] axis of the Pt-sputtered film. Figure 4(b) indicates that the Pt(100) film was hetero-epitaxially grown on the MgO(100) wafer. Figure 4(c) indicates that spot of Pt and BST superimposed on each other due to the same preferred orientation of (100). Figure 4(d) indicates that the orientation of the hetero-epitaxially grown $(Ba_{0.7}Sr_{0.3})TiO_3$ thin film was the (100) direction. The $(Ba_{0.7}Sr_{0.3})TiO_3$ thin film was highly hetero-epitaxial with respect to the Pt-sputtered film. These results indicated that the (100) three-axis-oriented the $(Ba_{0.7}Sr_{0.3})TiO_3$ thin film was deposited on the Pt(100)/MgO(100) substrate. These results agreed with the results of the XRD pattern and X-ray pole figure measurement.

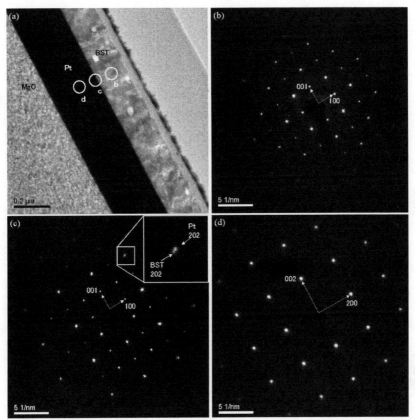

Fig. 4. (a) Cross-sectional TEM images of the BST thin film heat treated at 800 °C. (b) SAED pattern from the Pt-sputtered film. (c) The SAED pattern from the boundary of the Pt-sputtered film and the BST thin film. In set shows Pt diffraction spot and BST diffraction spot of (202). (d) The SAED pattern from the BST thin film.

Figure 5 shows a cross-sectional bright field TEM image of the boundary of the Pt-sputtered film and the $(Ba_{0.7}Sr_{0.3})TiO_3$ thin film heat treated at 800 °C. The $(Ba_{0.7}Sr_{0.3})TiO_3$ thin film was highly hetero-epitaxial with respect to the Pt-sputtered film, and was effectively single-crystal-like with no visible grain boundaries. The TEM observation results revealed that the $(Ba_{0.7}Sr_{0.3})TiO_3$ thin film heat treated at 800 °C exhibited a strong cube-on-cube epitaxy.

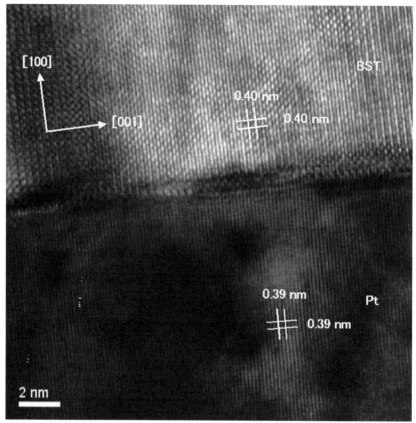

Fig. 5. Cross-sectional bright field TEM image of the boundary of the Pt-sputtered film and the BST thin film heat treated at 800 °C.

Figure 6 shows a C-V plot for the films at a frequency of 1 kHz. The measurement of the C-V plot in the MDM configuration revealed the dielectric properties. Two maximums were clearly observed in Figure 6. Butterfly loops were an indication of non linear dielectricity in the film. At zero bias, the dielectric constant and tan delta of the film heat treated at 650 °C were approximately 600 and 0.036, and dielectric constant and tan delta of the film heat treated at 800 °C is around 1200 and 0.022, respectively. This indicated that the dielectric constant of the three-axis-oriented (Ba$_{0.7}$Sr$_{0.3}$)TiO$_3$ thin film was twice as high as that of the poly crystal (Ba$_{0.7}$Sr$_{0.3}$)TiO$_3$ thin film.

The dielectric constant of the (Ba$_{0.6}$Sr$_{0.4}$)TiO$_3$ film epitaxially grown on the Pt(100)/MgO(100) substrate by the CVD method was 1200, as mentioned in ref. 11. This indicated that identical electric properties were observed though the film deposition processes were different.

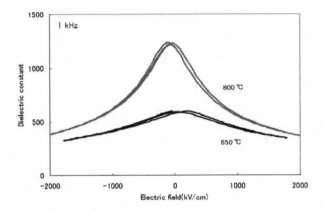

Fig. 6. C-V plots for the BST thin films heat treated at 650 °C and 800 °C at 1 kHz.

The temperature dependence of the dielectric constant in the films is shown in Fig. 7.The dielectric constant was measured at 1 kHz, with 0.1 V at zero-biased from -60 °C to 150 °C. Temperature dependence of the dielectric constant of the film heat treated at 650 °C was stable from -50 °C to 40 °C and decreased continuously with increasing temperature above 40 °C. This change at 40 °C is corresponded to some kind of a phase transition. The dielectric constant of the film heat treated at 800 °C constant decreased continuously from 1367 to 626 with increasing temperature.

In spite of, the CSD method being kinetically limited as compared with physical vapor deposition (PVD) method, (Ba$_{0.7}$Sr$_{0.3}$)TiO$_3$ three-axis-oriented thin films can be obtained by selecting optimum type of substrate and optimum conditions for epitaxial growth by the CSD method.

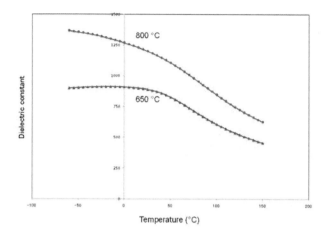

Fig. 7. The temperature dependence of the dielectric constant in the films film heat treated at 650 °C and 800 °C

IV. CONCLUSIONS

Perovskite (100)-three-axis-oriented (Ba₀.₇Sr₀.₃)TiO₃ thin films were hetero-epitaxially grown on a Pt(100)/MgO(100) substrate by the CSD method. The obtained perovskite (100) three-axis-oriented (Ba₀.₇Sr₀.₃)TiO₃ thin films were characterized by SEM, XRD, X-ray pole figure measurement, TEM, and SAED. The morphology and the crystallinity of thin films can be controlled by varying the heat treating conditions. The C-V measurement and the temperature dependence of the dielectric constant revealed that the electrical properties are controlled by varying the heat treating conditions.

Moreover, the fabrication of the thin films using the present method will be useful for the production of thin film devices using (Ba₀.₇Sr₀.₃)TiO₃.

REFERENCES
[1]Yutaka Takeshima, Kosuke Shiratsuyu, Hiroshi Takagi and Yukio Sakabe, "Preparation and Dielectric Properties of the Multilayer Capacitor with (Ba,Sr)TiO₃ Thin Layers by Metalorganic Chemical Vapor Deposition." Jpn. J. Appl. Phys. Vol. 36(1997) 5870
[2]C. M. Chu and P. Lin, "Electrical properties and crystal structures of (Ba,Sr)TiO₃ films and BaRuO₃ bottom electrodes prepared by sputtering." Appl. Phys. Lett. 70 [2] 249-51 (1997)
[3]D. Y. Wang, J. Wang, H. L. W. Chan, and C. L. Choy, "Structural and electro-optic properties of Ba₀.₇Sr₀.₃TiO₃ thin films grown on various substrates using pulsed laser deposition." J. Appl. Phys. 101, 043515 (2007)
[4]Osamu Nakagawara, Toru Shimuta, Takahiro Makino, Seiichi Arai, Hitoshi Tabata, and Tomoji Kawai, "Epitaxial growth and dielectric properties of (111) oriented BaTiO₃/SrTiO₃ superlattices by pulsed-laser deposition." Appl. Phys. Lett. 77, 3257 (2000)
[5]Shinichi Ito, Hiroshi Funakubo, Ivoyl P. Koutsaroff, Marina Zelner, and Andrew Cervin-Lawry, "Effect of the thermal expansion matching on the dielectric tenability of (100)-one-axis-oriented (Ba₀.₅Sr₀.₅)TiO₃ thin films." Appl. Phys. Lett. 90, 142910 (2007)
[6]H. Funakubo, Y. Takeshima, D. Nagano, K. Shinozaki, N. Mizutani, "Crystal structure and dielectric property of epitaxially grown (Ba,Sr)TiO₃ thin film prepared by molecular chemical vapor deposition." J. Mater. Res. Vol. 13, No. 12 (1998) 3512-3518
[7]Jian-Gong Cheng, Jun Tang, Xiang-Jian Meng, Shao-Ling Guo, Jun-Hao Chu, Min Wang, Hong Wang,

Zhuo Wang. "Fabrication and Characterization of Pyroelectric $Ba_{0.8}Sr_{0.2}TiO_3$ Thin Films by a Sol-Gel Process." Journal of the American Ceramic Society 84 (7), 1421–1424 (2001)

[8]Danielle M. Tahan, Ahmad Safari, Lisa C. Klein, "Preparation and Characterization of $Ba_xSr_{1-x}TiO_3$ Thin Films by a Sol-Gel Technique." Journal of the American Ceramic Society 79 (6), (1996) 1593–1598

[9]S. Hoffmann, R. Waser, "Control of the morphology of CSD-prepared $(Ba,Sr)TiO_3$ thin films." Journal of the European Ceramic Society Volume 19, Issues 6-7, June 1999, Pages 1339-1343

[10]Tadasu Hosokura, Akira Ando and Yukio Sakabe, "Fabrication and Electrical Characterization of Epitaxially Grown $(Ba,Sr)TiO_3$ Thin Films Prepared by Sol-Gel Method." Key Engineering Materials Vol. 320 (2006) pp. 81-84

[11]Yutaka Takeshima, Katsuhiko Tanaka and Yukio Sakabe, "Thickness Dependence of Characteristics for $(Ba,Sr)TiO_3$ Thin Films Prepared by Metalorganic Chemical Vapor Deposition." Jpn. J. Appl. Phys. Vol.39 (2000) 5389-5392

INFLUENCE OF Ca CONCENTRATION IN $(Ba,Ca)TiO_3$ BASED CERAMICS ON THE RELIABILITY OF MLCCs WITH Ni ELECTRODES

Jun Ikeda, Shoichiro Suzuki, Toshikazu Takeda, Akira Ando and Hiroshi Takagi
Murata Mfg. Co. Ltd. 1-10-1 Higashikotari, Nagaokakyo-shi, Kyoto 617-8555, Japan

ABSTRACT
The influence of Ca concentration in $(Ba,Ca)TiO_3$ based ceramics on the reliability of multilayer ceramic capacitor with Ni electrodes was investigated. We prepared $(Ba_{1-x}Ca_x)TiO_3$ based ceramics where x is from 0 to 0.2. The microstructure analyses were carried out with Scanning electron microscope, X-ray diffraction, and electron probe microanalysis. The reliability which estimated by a highly accelerated life test from 150°C to 175°C increased with Ca concentration up to x=0.12. The activation energy of the electrical degradation increased with increasing Ca concentration until x = 0.10. We can interpret that this result originated from a low mobility of oxygen vacancy caused by lattice shrinkage with Ca ion substitution to Ba site. On the other hand, the reliability decreased where x is above 0.12. The microstructure analyses showed $CaTiO_3$ segregation phase appeared at x above 0.12. The reliability is thought to decrease corresponding to increase $CaTiO_3$ segregation phase.

INTRODUCTION
Multilayer ceramic capacitors (MLCCs) are widely used in advanced electric devices because of their high volumetric efficient and superior performance at high frequency [1]. Increasing the layer count of the MLCCs using thin and high dielectric constant materials has resulted in their miniaturization and the maximization of the capacitance of the MLCCs. However, an increasing in the number of layers leads to an increase of the electrode cost, hence the internal electrodes used have been changed from those of conventional noble metals to those of based metals.

The $BaTiO_3$ based MLCCs with Ni internal electrodes must be fired in a reducing atmosphere to avoid oxidation of the internal electrodes. If the oxygen partial pressure, $P(O_2)$, is not sufficiently low, the Ni internal electrodes are partially oxidized and diffuse into the ceramic, resulting in a poor insulation resistance. Therefore, the sintering atmosphere must be strictly controlled to maintain the $P(O_2)$ lower than that required for the Ni/NiO equilibrium. As a result, there is a large quantity of oxygen vacancy in the $BaTiO_3$ active layers. The reliability of MLCCs with Ni internal electrodes is thought to be dominated by electromigration of oxygen vacancy in the dielectric layers [2, 3].

Recently, the requirements of much higher performance such as miniaturization and high capacitance are increased and the thickness of active layer has been reduced to satisfy these requirements. Moreover, the recent automotive applications have demanded higher reliability in MLCCs and their dielectrics. Therefore, improving reliability of dielectrics become more important. Sakabe et al reported that $(Ba,Ca)TiO_3$ based ceramics show higher reliability and higher resistivity than $BaTiO_3$ based ceramics sintered in a reducing atmosphere [4, 5]. Therefore, we have been interested in $(Ba,Ca)TiO_3$ based ceramics properties and investigating them. For example, Takeda et al investigated about the influence of Ca concentration on the dielectric properties of $(Ba,Ca)TiO_3$ based ceramics [6]. However, Ca concentration dependence to the reliability and the resistivity has not been fully clarified yet. In this study, therefore, the influence of Ca concentration in $(Ba,Ca)TiO_3$ based ceramics on the reliability and the resistivity was investigated.

EXPERIMENTAL

Ca doped BaTiO$_3$ powders (Ba$_{1-x}$Ca$_x$)TiO$_3$ were synthesized by conventional powder processing using BaCO$_3$, CaCO$_3$ and TiO$_2$ as starting materials, where x varied from 0 to 0.2. They were mixed using ball milling and calcined at 1100°C in air, in order to obtain (Ba$_{1-x}$Ca$_x$)TiO$_3$ powders. Lattice parameters were measured by X-ray diffraction (XRD). Y$_2$O$_3$, MgO, MnO$_2$ and SiO$_2$ were added to the (Ba$_{1-x}$Ca$_x$)TiO$_3$ powder. MLCCs samples were prepared the following procedure. Formulated (Ba$_{1-x}$Ca$_x$)TiO$_3$ powders were mixed well with organic binder, and solvent using ball milling in order to prepare the homogeneous slurry. Green sheets were formed by doctor-blade casting process and the thickness was about 3.5 microns. After casting and drying, Ni electrodes were printed on the sheets using screen-printing method. These sheets were stacked, pressed, and cut into green MLCCs samples. After binder burnout, the green samples were fired at temperature between 1200°C and 1300°C for 2hour under reducing atmosphere between 1*10^{-9} and 1*10^{-10} MPa of O$_2$ partial pressure controlled by N$_2$-H$_2$-H$_2$O.

Scanning electron microscope (SEM) analyses were performed to measure the particle size of calcined powders and grain size of ceramics. Distribution of elements in the ceramics was measured by electron probe microanalysis (EPMA). The crystal phases was identified by X-ray diffraction (XRD) at room temperature.

The reliability of MLCCs was evaluated by the highly accelerated life test (HALT), measuring the insulation resistance (IR) at 150°C, 160°C and 175°C under applied voltage of 30kV/mm. Mean time to failure (MTTF) was obtained from the Weibull plot analysis of the failure time of the samples. Electric current-electric field (J-E) characteristic were conducted at 175°C.

RESULTS AND DISCUSSION
Characterization of powders

The crystal structure and lattice parameters of $(Ba_{1-x}Ca_x)TiO_3$ powders were determined by XRD analyses. Figure 1 shows XRD patterns of the $(Ba_{1-x}Ca_x)TiO_3$ powders. The $(Ba_{1-x}Ca_x)TiO_3$ powders had single phase and showed a tetragonal perovskite structure where x is from 0 to 0.14, while where x is above 0.16, CaTiO$_3$ segregation phase slightly appeared. The lattice parameters were obtained from XRD patterns. Figure 2 shows the relationship between Ca concentration and lattice volume of the $(Ba_{1-x}Ca_x)TiO_3$ powders at room temperature. The lattice volume monotonously decreased with increasing Ca concentration. This result agrees well with former works by Mitsui et al [7] and Sakabe et al [5]. It is explained by the smaller ionic radius of Ca substituted to Ba site in BaTiO$_3$ crystal.

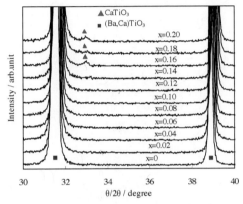

Figure 1. XRD patterns of $(Ba_{1-x}Ca_x)TiO_3$ powders at room temperature.

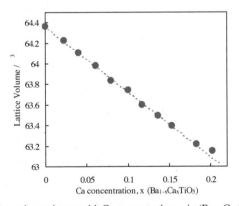

Figure 2. Lattice volume change with Ca concentration, x in $(Ba_{1-x}Ca_x)TiO_3$ powders.

Microstructures

An SEM micrograph of (Ba$_{0.9}$Ca$_{0.1}$)TiO$_3$ powders is shown in Figure 3(a). Ceramic samples of 92 to 95% theoretical density were obtained under sintering higher than 1250°C. The average grain sizes of ceramics sintered at 1250°C were about 250nm as shown Figure 3(b). It is confirmed that the grain growth was not eminent during sintering.

We identified the secondary phases in the crashed MLCCs samples by XRD at room temperature. Figure 4 shows XRD patterns where Ca concentration, x is from 0 to 0.20. Y$_2$O$_3$ phase and Y$_2$Ti$_2$O$_7$ phase were detected where x is from 0 to 0.10, while where x is above 0.10, Y$_2$O$_3$ phase and Y$_2$Ti$_2$O$_7$ phase were disappeared and CaTiO$_3$ phase was slightly detected.

Figure 5 shows the compositional distribution in dielectric layers of the MLCCs samples analyzed by EPMA. Where x is above 0.10, highly concentration spots of Ca were observed and increasing with Ca concentration of (Ba$_{1-x}$Ca$_x$)TiO$_3$. XRD and EPMA results revealed that the microstructures were almost same where x is from 0 to 0.10, while segregation phases changed to CaTiO$_3$ where x is more than 0.12

Figure 3. SEM micrograph of (Ba$_{0.9}$Ca$_{0.1}$)TiO$_3$ (a)powders and (b)ceramics

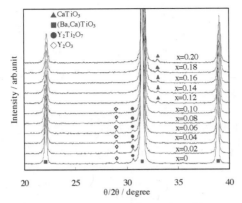

Figure 4. XRD patterns of crashed MLCCs sample for various Ca concentration's $(Ba_{1-x}Ca_x)TiO_3$ based ceramics.

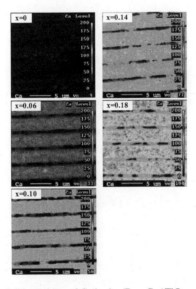

Figure 5. Distribution of Ca in the $(Ba_{1-x}Ca_x)TiO_3$ ceramics

Resistivity at high temperature

Because the reliability of the dielectrics expected to be related with electrical conduction, J-E characteristics were measured at 175°C for the MLCCs samples with various Ca concentrations to understand the electrical conduction and results are shown in Figure 6(a). The doping Ca tended to increase the leakage current at low electric field, while at high electric field, the leakage current did not change so much with Ca concentration. Figure 6(b) shows the slopes of logJ-logE plot vs. logE. The slopes of high electric field were decreased with increasing Ca concentration until where x is 0.12, while x is above 0.12, the slopes did not change with Ca concentration. Increasing the slopes means that current increased more abruptly and electrical conduction has changed under high voltage. Therefore, this result is supposed to mean that electrical conduction is suppressed with increasing Ca concentration under high voltage.

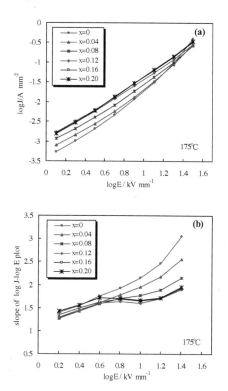

Figure 6. E-J characteristic at 175°C (a)logJ vs logE, (b)slope of logJ-logE plot vs. logE

Reliability of (Ba$_{1-x}$Ca$_x$)TiO$_3$ ceramics

Figure 7 shows an MTTF of HALT under the DC field strength 30kV/mm at 175°C against Ca concentration. The MTTF increased to 20h with increasing Ca concentration until x=0.12, while decreased at x above 0.12. It was confirmed that degradation behaviors of the dielectrics depend on the Ca concentration in these harsh conditions.

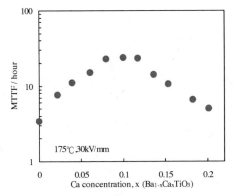

Figure 7. The mean time to failure change with Ca concentration at highly accelerated life test.

In order to clarify the activation energy of electrical degradation, the HALT were conducted at 150°C, 160°C, and 175°C. According to the empirical equation modeled by Prokopowicz and Vaskas[8], the MTTF under the HALT is expressed by

$$\frac{t_1}{t_2} = \left(\frac{V_2}{V_1}\right)^n \exp\left[\frac{Ea}{k}\left(\frac{1}{T_1} - \frac{1}{T_2}\right)\right] \tag{1}$$

, where t is the MTTF, V is the applied dc voltage, and T is the absolute temperature. Ea is the activation energy of electrical degradation, k is Boltzmann's constant, n is the voltage acceleration factor and subscripts refer to two different test conditions. Figure 8 shows the activation energy change against Ca concentration. The activation energy increased with increasing Ca concentration until x = 0.10 and decreased where x is above 0.10.

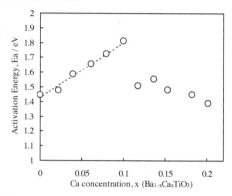

Figure 8. The relationship between Ca concentration and the activation energy from MTTF at HALT

The degradation of insulation resistance observed during the HALT is due to the electromigration of oxygen vacancy and locally concentrated oxygen vacancy near the cathode may lead to the leakage current increase [2, 3]. As shown in Figure 7, the MTTF was improved with increasing Ca concentration until x=0.12. This tendency is similar to that of electrical conduction shown in Figure 6(b) where x is less than 0.12. We would expect that the change of electrical conduction under high voltage related with the electrical degradation.

For improving the reliability, it is necessary to decrease an amount of oxygen vacancy or to decrease a mobility of oxygen vacancy. While Sakabe et al[5] and Suzuki et al[9] reported that Ca doping in the BaTiO₃ increases the amount of oxygen vacancy, it is therefore expected that the reliability depends on the mobility of oxygen vacancy, not on its amount. As shown in Figure 2, Ca doping at Ba site make lattice of BaTiO₃ shrink. Also, we supposed that the lattice shrinkage might make Ti-O bonding strong. Therefore, we would expect this result as being due to suppress the rate of electromigration of oxygen vacancy caused by lattice shrinkage with Ca at the Ba site. Moreover, the change of electrical conduction in Figure 6(b) might originate from suppressing oxygen ion conduction with increasing Ca concentration. The relationship between doping Ca to BaTiO₃ and Ti-O bonding energy would be investigated in future works.

On the other hand, for x more than 0.12, the MTTF is thought to decrease corresponding to the increasing CaTiO₃ segregation phase. Under the coexistence of the segregation phase, the degradation behavior cannot be explained only by the Ca concentration in ceramic grain.

CONCLUSION

The influence of Ca concentration in (Ba,Ca)TiO₃ based ceramics on the reliability of multilayer ceramic capacitor with Ni electrode was investigated. The reliability which estimated by HALT from 150°C to 175°C increased with Ca concentration up to x=0.12. The activation energy of degradation behavior increased with increasing Ca concentration until x=0.10. The lattice volume of $(Ba_{1-x}Ca_x)TiO_3$ monotonously decreased with increasing Ca concentration. We can interpret this result originated from a low mobility of oxygen vacancy caused by lattice shrinkage with Ca ion substitution to Ba site.

On the other hand, the reliability decreased where x is above 0.12. The microstructure analyses showed CaTiO₃ segregation phase appeared at x above 0.12. The MTTF is thought to decrease corresponding to the increasing CaTiO₃ segregation phase.

Appling this technology, (Ba,Ca)TiO₃ based ceramics are developed to realize high performance MLCCs with Ni electrode.

REFERENCES

[1]K. Wakino, T. Sato, T. Ushiro, and Y. Sakabe, Proc. 3rd Capacitor and Resistor Technology Symp., 183 (1983).

[2]R. Waser, Electrochemical Boundary Conditions for Resistance Degradation of Doped Alkaline-Earth Titanates, *J. Am. Ceram. Soc.*,72, 2234-40 (1989).

[3]T. Baiatu, R. Waser, and K. H. Haerdtl, dc Electrical Degradation of Perovskite-Type Titanates: A Model of the Mechanism, *J. Am. Ceram. Soc.* 73, 1663-73 (1990).

[4]Y. Sakabe, K. Minai, and K. Wakino, High-Dielectric Constant Ceramics for Base Metal Monolithic Capacitor, *Jpn. J. Appl. Phys. Suppl*, 20, 147-50 (1981).

[5]Y. Sakabe, N. Wada, H. Hiramatsu, and T. Tonogaki., Dielectric Properties of Fine-Grained BaTiO₃ Ceramics Doped with CaO, *Jpn. J. Appl. Phys.*, 41, 6922-25 (2002).

[6]T. Takeda, H. Sano, T. Morimoto, and H. Takagi, The Influence of Ca Concentration on the Dielectric Properties of (Ba,Ca)TiO₃ Based Ceramics, 13th US-Japan Seminar on Dielectric and Piezoelectric Ceramic, Extended Abstract, 343-46 (2007).

[7]T. Mitsui and W. B. Westphal, Dielectric and X-Ray Studies of $Ca_xBa_{1-x}TiO_3$ and $Ca_xSr_{1-x}TiO_3$, *Phys. Rev.*, 124, 1354-59 (1961).

[8]T. I. Prokopowicz and A. R. Vaskas, Research and Development Intrinsic Reliability Subminiature Ceramic Capacitors, NTIS, AD-864068, (1969).

[9]T. Suzuki, M. Ueno, Y. Nishi, and M. Fujimoto, Dislocation Loop Formation in Nonstoichiometric (Ba,Ca)TiO₃ and BaTiO₃ Ceramics, *J. Am. Ceram. Soc.*, 84, 200-206 (2001).

CRYSTAL STRUCTURE DEPENDENCE OF ELECTRICAL PROPERTIES OF
$Li_{0.02}(K_{1-x}Na_x)_{0.98}NbO_3$ CERAMICS

Eung Soo Kim[1], Seock No Seo[1], Jeong Ho Cho[2] and Byung Ik Kim[2]
[1]Department of Materials Engineering, Kyonggi University, Suwon 443-760, Korea
[2]Korea Institute of Ceramic Engineering and Technology, Seoul 153-023, Korea

ABSTRACT

Effects of crystal structure on the electrical properties of $Li_{0.02}(K_{1-x}Na_x)_{0.98}NbO_3$ ($0.4 \leq x \leq 0.6$) ceramics have been investigated. For the specimens sintered at 1030°C for 5 hr, the unit-cell volume was decreased with the increase of $NaNbO_3$ content. This result could be attributed to the smaller ionic radius of Na^+(0.139 nm) than that of K^+(0.164 nm). Dielectric constant and electromechanical coupling factor (k_p) were dependent on the Nb-site bond valences. The mechanical quality factor (Q_m) of the sintered specimen was maximum value of 229 at $x=0.5$, and then decreased due to the decrease of relative density. The effect of octahedral distortion on the temperature coefficient of k_p (TCk_p) for $Li_{0.02}(K_{1-x}Na_x)_{0.98}NbO_3$ perovskite structure was also discussed.

INTRODUCTION

Recently, much attention has been paid to the lead-free piezoelectric ceramics because the toxicity of lead can contaminate the environment and damage human health. Potassium sodium niobate $(K_{1-x}Na_x)NbO_3$ (KNN) based ceramics are the most promising candidate for lead-free piezoelectric ceramics due to the high Curie temperature, the good electrical properties and environmental compatibility, comparing to other candidates such as bismuth sodium titanate $(Bi_{1-x}Na_x)TiO_3$ and bismuth titanate $(Bi_4Ti_3O_{12})$.[1, 2]

Various attempts have been tried to obtain the well sintered specimens with good electrical properties,[3, 4] and structural characteristics of KNN based ceramics were reported.[5, 6] However, a major problem of KNN is reported that it is not easy to obtain the well sintered specimens with high density by conventional sintering in air, and the crystal structural nature of KNN is presently not well understood, especially near a virtual morphotropic phase boundary. Electrical properties of materials are strongly dependent on the composition of materials, the chemical nature of constituent ions, the distances between cations and anions and the structural characteristics originating from the bonding type which can be evaluated by the bond valence.[7] The fundamental relationships between the structural characteristics and the piezoelectric properties also have to be known to control and improve the piezoelectric properties of materials effectively.

Therefore, this study has been focused to investigate the electrical properties of $Li_{0.02}(K_{1-x}Na_x)_{0.98}NbO_3$ with 0.5wt% ZnO ($0.4 \leq x \leq 0.6$) ceramics based on the structural characteristics in view point of Nb-site bond valence and oxygen octahedral distortion.

EXPERIMENTAL PROCEDURE

The compositions of $Li_{0.02}(K_{1-x}Na_x)_{0.98}NbO_3$ with 0.5wt% ZnO ($0.4 \leq x \leq 0.6$) were prepared by the conventional mixed oxide method. Li_2CO_3, K_2CO_3, Na_2CO_3 and Nb_2O_5 powders with high-purity (99.9%) were used as the starting materials. They were milled using ZrO_2 balls for 24h in ethanol and then dried. The dried powders were calcined from 800°C for 5h, and then milled again with 0.5wt% ZnO for 24 h, to improve the sinterability of the specimens. The dried powders were pressed into 10mm diameter disk at 1500 kg/cm^2 isostatically. These pellets were sintered at 1030°C for 5h in air.

Powder X-ray diffraction analysis (D/Max-3C, Rigaku, Japan) with Cu Kα radiation was used to determine the crystalline phases in the calcined and the sintered specimens. XRD data for a Rietveld analysis were collected over a range of $2\theta=20°\sim80°$ with a step size of 0.02° and a count time of 2s. From Rietveld refinements of the X-ray diffraction patterns using the Fullprof program[8], the lattice parameters and atomic positions of the sintered specimens were determined. The interactive crystallography software package CrystalMaker (2.1)[9] was used to simulate the NbO_6 octahedron form crystallographic data, and the individual bond lengths was determined. The polished surface of the sintered specimens was observed using a scanning electron microscope (JEOL JSM-6500F, Japan).

Silver electrodes were formed on both surfaces of each sintered disk by firing at 700°C for 10 min. The samples were polarized in silicon oil bath at 80°C by applying a DC electric field (5 kV/mm for 15 min). The electromechanical coupling coefficient (k_p) and mechanical quality factor (Q_m) was determined by the resonance and anti-resonance method on the basis of IEEE standards[10] using an impedance analyzer (HP 4192A, Palo Alto, CA, USA).

RESULTS AND DISCUSSION

3.1 Crystalline Structure and Physical Properties

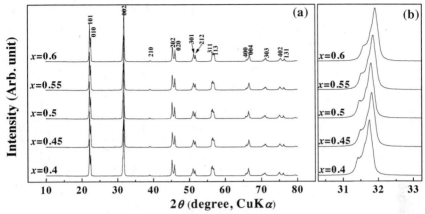

Figure 1. X-ray diffraction patterns of Li$_{0.02}$(K$_{1-x}$Na$_x$)$_{0.98}$NbO$_3$ with 0.5wt% ZnO (0.4<x<0.6) ceramics sintered at 1030°C for 5h.

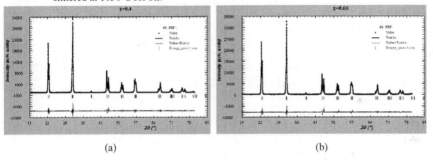

Figure 2. Observed and calculated X-ray diffraction patterns by Rietveld analysis for Li$_{0.02}$(K$_{1-x}$ Na$_x$)$_{0.98}$NbO$_3$ with 0.5wt% ZnO sintered specimens : (a) x=0.4, (b) x=0.6

Table 1. Lattice Parameters and Refinement Statistics obtained from Rietveld Refinement for Li$_{0.02}$(K$_{1-x}$Na$_x$)$_{0.98}$NbO$_3$ with 0.5wt% ZnO Sintered Specimens.

x (mol)	a (Å)	b (Å)	c (Å)		atomic positions				V$_{unit-cell}$ /Z (Å^3)	Rwp (%)	Rp (%)
					Na/K	Nb	O$_1$	O$_2$			
				x	0.0	0.0	0.0	0.20954(1)			
0.4	5.676(2)	3.946(1)	5.645(1)	y	0.5	0.0	0.5	0.0	63.223	12.9	9.11
				z	0.49011(0)	0.0	0.01048(2)	0.24074(0)			
				x	0.0	0.0	0.0	0.29315(0)			
0.45	5.672(2)	3.943(1)	5.640(0)	y	0.5	0.0	0.5	0.0	63.080	12.5	8.65
				z	0.50871(1)	0.0	0.01181(1)	0.26101(0)			
				x	0.0	0.0	0.0	0.29111(0)			
0.5	5.666(1)	3.940(1)	5.634(2)	y	0.5	0.0	0.5	0.0	62.882	12.4	8.71
				z	0.51524(0)	0.0	0.00811(0)	0.25987(1)			
				x	0.0	0.0	0.0	0.29432(0)			
0.55	5.661(0)	3.938(2)	5.628(1)	y	0.5	0.0	0.5	0.0	62.726	12.8	9.06
				z	0.50751(1)	0.0	0.01318(2)	0.26433(0)			
				x	0.0	0.0	0.0	0.30069(0)			
0.6	5.656(0)	3.934(1)	5.622(3)	y	0.5	0.0	0.5	0.0	62.528	12.0	8.36
				z	0.51271(0)	0.0	0.02100(0)	0.26101(1)			

Figure 1 showed the X-ray diffraction (XRD) patterns of Li$_{0.02}$(K$_{1-x}$Na$_x$)$_{0.98}$NbO$_3$ with 0.5wt%

ZnO ($0.4 \leq x \leq 0.6$) sintered at 1030°C for 5h in air. For the entire range of compositions, the single phase with orthorhombic perovskite structure was detected, and the (002) peak positions of XRD patterns shifted to higher 2θ angle with the increase of NaNbO$_3$ content (x), as shown in Fig.1(b). Figure 2 shows the Rietveld refinement analyses of X-ray diffraction patterns for Li$_{0.02}$ (K$_{1-x}$Na$_x$)$_{0.98}$NbO$_3$ with 0.5wt% ZnO at x=0.4 and x=0.6. For the refinement procedures, the initial structure model for orthorhombic perovskite compounds was taken from the study by L. Katz et al. (ICSD # 3813, Bmm2).[11] The lattice parameters (a, b and c-axis) and atomic positions obtained from Rietveld refinement for the specimens are listed in Table 1. With an increase of $NaNbO_3$ content, the unit cell volume of the specimens decreased, due to the smaller ionic radius of Na^+(0.139nm) than that of K^+(0.164 nm).[12] For the sintered specimens with space group B$mm2$, the coordination number of Nb-site is 6 and that of K/Na-site is 12, and NbO_6 octahedra are connected with other NbO_6 octahedra by edge sharing in the unit cell. The NbO_6 octahedra were composed of three types of bond length, and were slightly distorted, as shown in Fig 3.

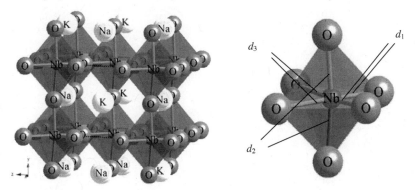

Figure 3. Crystal structure (B$mm2$) of $Li_{0.02}(K_{1-x}Na_x)_{0.98}NbO_3$ with 0.5wt% ZnO ceramics sintered at 1030°C for 5h.

Although the ions are incorporated in the same structure, their bond length and strength are changed due to the interactions with their surrounding ions. These interactions lead to the crystal structural modifications and determine the physical properties of materials. Relationships between ionic interactions and bond length can be evaluated quantitatively by the bond valence, which is the actual valence of the ions affected by the surrounding ions in the crystal structure. Hence, the bond length is a unique function of bond valence. The bond valences between Nb^{5+} ion and oxygen ion were calculated by equations (1) and (2).[7]

$$V_i = \Sigma_j v_{ij} \tag{1}$$

$$v_{ij} = \exp\left\{\frac{(R_{ij} - d_{ij})}{b}\right\} \tag{2}$$

where R_{ij} is the bond valence parameter, d_{ij} is the length of a bond between atom i and j, and b is commonly taken to be a universal constant equal to 0.37 Å. Additionally, the NbO_6 octahedral distortion calculated from individual bond lengths[12] was closely related with Nb-site bond valence (V_{Nb}). The octahedral distortion was increased with increase of V_{Nb}. These individual bond lengths could be changed by substitution of $LiNbO_3$ and $NaNbO_3$ for $KNbO_3$. However, a $LiNbO_3$ content of the specimens was same value (2 molar %) through the entire range of compositions. Therefore, the changes of Nb-site bond valence (V_{Nb}) and octahedral distortion were only dependent on the $NaNbO_3$ content, as listed in Table 2.

Table 2. Crystallographic Data, Nb-site Bond Valence and Oxygen Octahedral Distortion of $Li_{0.02}(K_{1-x}Na_x)_{0.98}NbO_3$ with 0.5wt% ZnO Sintered Specimens.

Space Group	x (mol)	d_1 (Å)	d_2 (Å)	d_3 (Å)	B-site Bond Valence (V_{Nb})	Octahedral Distortion($\Delta \times 10^4$)	Unit-cell Distortion (a/b)
	0.4	1.811	1.976	2.197	5.225	62.65	1.4386
	0.45	1.795	1.974	2.209	5.319	72.41	1.4385
Bmm2	0.5	1.799	1.964	2.215	5.321	73.69	1.4380
	0.55	1.781	1.968	2.209	5.450	77.82	1.4378
	0.6	1.776	1.968	2.219	5.466	83.10	1.4377

Figure 4 shows SEM micrographs of the sintered specimens. Due to the lower sintering temperature at $1030°C$ for 5h, a liquid phase was not appeared during sintering process, and each composition showed well-grown grains and less porosity. The grain size of the specimen with $x=0.5$ was slightly larger than that of the other specimens, and the relative densities of the specimen with $x=0.5$ showed also maximum value. As shown in Fig. 5(a), the relative density was increased with $NaNbO_3$ content up to $x=0.5$, and then decreased. This result could be affected to mechanical quality factor (Q_m). As shown in Fig. 5(b), the quality factor (Q_m) of the specimen was also shown maximum value at $x=0.5$, due to the highest value of relative density.

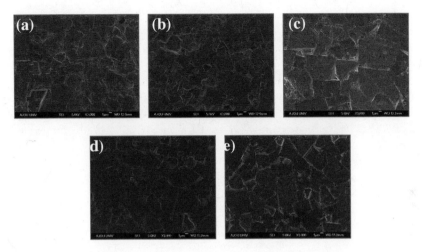

Figure 4. Scanning electron microscopic micrograph of $Li_{0.02}(K_{1-x}Na_x)_{0.98}NbO_3$ with 0.5wt% ZnO (0.4<x<0.6) ceramics sintered at 1030°C for 5h : (a) $x = 0.4$, (b) $x = 0.45$, (c) $x = 0.5$, (d) $x = 0.55$, (e) $x = 0.6$.

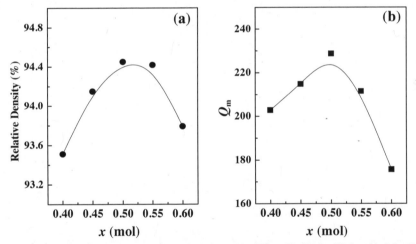

Figure 5. Relative density (a) and quality factor (Q_m) (b) of $Li_{0.02}(K_{1-x}Na_x)_{0.98}NbO_3$ with 0.5wt% ZnO (0.4 <x <0.6) ceramics sintered at 1030°C for 5h.

3.2 Electrical Properties

It has been reported that the electrical properties of pure $(K_{1-x}Na_x)NbO_3$ system was affected by the morphotropic phase boundary(MPB).[13] However, only a orthorhombic phase was detected for the specimens of $Li_{0.02}(K_{1-x}Na_x)_{0.98}NbO_3$ with 0.5wt% ZnO sintered at 1030°C for 5h through the entire composition range $(0.4 \leq x \leq 0.6)$ in this study.

As shown in Fig. 6(a), the dielectric constant of the specimens decreased with increasing of $NaNbO_3$ content, and this result is agreed well with the reports[14] for $(K_{1-x}Na_x)NbO_3$ system. This result could explained that the dielectric constant was affected by the rattling effect of center ions in the NbO_6 octahedra[15] which was closely related with bond length and bond strength between Nb^{5+} to O^{2-}. These structural characteristics (bond length and bond strength) could be evaluated by bond valence. Due to the decreasing of bond lengths between Nb^{5+} and O^{2-}, the Nb-site bond valence (V_{Nb}) increased with increasing of $NaNbO_3$ content (Fig. 6(b)), which in turn, the dielectric constant decreased with $NaNbO_3$ content. However, the reason is not clear why the electromechanical coupling factor (k_p) of the specimens increased with $NaNbO_3$ content and it is under investigated at present. It may be also associated with the increase of Nb-site bond valence.

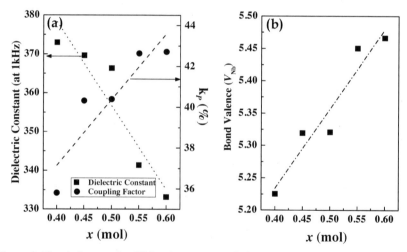

Figure 6. Electrical properties (dielectric constant and electromechanical coupling factor) (a) and bond valence (V_{Nb}) (b) of $Li_{0.02}(K_{1-x}Na_x)_{0.98}NbO_3$ with 0.5wt% ZnO $(0.4<x<0.6)$ ceramics sintered at 1030°C for 5h.

Figure 7(a) shows the dependence of electromechanical coupling factor (k_p) on the temperature. With the increasing of temperature, the coupling factor (k_p) slightly increased up to around 170°C (depends on $NaNbO_3$ content from 170.2°C to 173.4°C), and then remarkably decreased at transition temperature (T_{O-T}) from orthorhombic to tetragonal. To investigate thermal stability of the specimen, the temperature coefficient of k_p (TCk_p) from 30°C to T_{O-T} was calculated from equation (3).[16]

$$TCk_p = \frac{k_{p,T_{O-T}} - k_{p,30°C}}{k_{p,30°C}} \qquad (3)$$

With increasing of $NaNbO_3$ content, the TCk_p of the specimens was increased and it was related to the structural symmetry which could be evaluated by the ratio of lattice parameter of a to b. For the $(K_{1-x}Na_x)NbO_3$ perovskite structure, the structural symmetry was increased with increasing of temperature.[17] The structural symmetry of the specimens could be evaluated by NbO_6 octahedral distortion and unit-cell distortion (the ratio of lattice parameter of a to b), which was changed with $NaNbO_3$ content, as shown in Table 2. Fig. 7(b) showed the dependence of TCk_p and unit-cell distortion on NbO_6 octahedral distortion for the specimens. With increasing of $NaNbO_3$ content, the octahedral distortion was increased, and the crystal structure turned to higher symmetrical structure due to the smaller unit-cell distortion. Finally the thermal stability of structure was decreased with $NaNbO_3$ content.

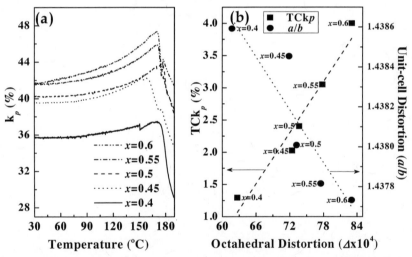

Figure 7. Temperature dependence of k_p (a) and octahedral distortion dependence of TCk_p (b) for $Li_{0.02}(K_{1-x}Na_x)_{0.98}NbO_3$ with 0.5wt% ZnO ($0.4<x<0.6$) ceramics sintered at 1030°C for 5h.

CONCLUSIONS

A single phase with orthorhombic perovskite structure was obtained through the entire composition range of $Li_{0.02}(K_{1-x}Na_x)NbO_3$ $(0.4 \leq x \leq 0.6)$ with 0.5wt% ZnO ceramics. From the Rietveld refinement of X-ray diffraction patterns, the orthorhombic perovskite ($Bmm2$, No.38) structure was detected. The quality factor (Q_m) of the specimens was increase up to $x=0.5$, and then decreased with $NaNbO_3$ content, due to the decrease of relative density. With the $NaNbO_3$ content, the dielectric constant and electromechanical coupling factor (k_p) was affected by Nb-site bond valence. Temperature coefficient of k_p (TCk_p) was dependent on octahedral distortion. The good piezoelectric properties of k_p= 42.5%, Q_m= 229 were obtained from $Li_{0.02}(K_{0.5}Na_{0.5})NbO_3$ with 0.5wt% ZnO ceramics sintered at 1030°C for 5h.

ACKNOWLEDGEMENT

This research was supported by a grant from the Fundamental R&D Program for Core Technology of Materials funded by the Ministry of Knowledge Economy, Republic of Korea.

REFERENCES

[1] R. E. Jaeger and L. Egerton, Hot Pressing of Potassium-Sodium Niobates, *J. Am. Ceram. Soc.*, **45**, 209-13 (1962).

[2] Y. Saito, H. Takao, T. Tani, T. Nonoyama, K. Takatori, T. Homma, T Nagaya and M. Nakamura, Lead-free Piezoceramics, *Nature*, **432**, 84–7 (2004).

[3] J.-F. Li, K. Wang, B.-P. Zhang and M. Zhang, Ferroelectric and Piezoelectric Properties of Fine-grained $Na_{0.5}K_{0.5}NbO_3$ Lead-free Piezoelectric Ceramics Prepared by Spark Plasma Sintering, *J. Am. Ceram. Soc.*, **89**, 706–09 (2006).

[4] R.P. Wang, R.J. Xie, T. Sekiya and Y. Shimojo, Fabrication and Characterization of Potassium-Sodium Niobate Piezoelectric Ceramics by Spark–Plasma–Sintering Method, *Mater. Res. Bull.*, **39**, 1709–15 (2004).

[5] A. Reisman and F. Holtzberg, Phase Equilibria in the System K_2CO_3-Nb_2O_5 by the Method of Differential Thermal Analysis, *J. Am. Chem. Soc.*, 77, 2115-19 (1955).

[6] H. Birol, D. Damjanovic and N. Setter, Preparation and Characterization of $(K_{0.5}Na_{0.5})NbO_3$ Ceramics, *J. Euro. Ceram. Soc.*, **26** 861–66 (2006).

[7] N. E. Brese and M. O'Keefe, Bond-Valence Parameters for Solids, *Acta. Cryst.*, **B47**, 192-97 (1991).

[8] T. Roisnel, J. Rodriguez-Carvajal, WinPLOTR: A Windows Tool for Powder Diffraction Patterns Analysis , *Material Science Forum*, **378-381,** 118-23 (2001).

[9] D. C. Palmer, CrystalDiffract, version 2.1.0: Interactive Crystallography for MacOS, CrystalMaker Software, Oxford, 1999.

[10] IRE Standards on Piezoelectricity, *IEEE Trans. Ultraso. Ferro.*, **43**, 719-22 (1996).

[11] L. Katz and H. D. Megaw, The Structure of Potassium Niobate at Room Temperature: The Solution of a

Pseudo-Symmetric Structure by Fourier Methods, *Acta Cryst.*, **22**, 639-48 (1967).

[12]R. D. Shannon, Revised Effective Ionic Radii and Systematic Studies of Interatomic Distances in Halides and Chalcogenides, *Acta Cryst.*, **A3**, 751-67 (1976).

[13]G. Shirane, R. Newnham and R. Pepinsky, Dielectric Properties and Phase Transitions of $NaNbO_3$ and $(Na,K)NbO_3$, *Phys. Rev.*, **96**, 581-88 (1954).

[14]V. Lingwa, B. S. Semwal and N. S. Panwar, Dielectric properties of $Na_{1-x}K_xNbO_3$ in orthorhombic phase, *Bull. Mater. Sci.*, **26** [6] 619–25 (2003).

[15]E.S. Kim and B.S. Chun, Estimation of Microwave Dielectric Properties of $[(Na_{1/2}La_{1/2})_{1-x}$ $(Li_{1/2}Nd_{1/2})_x]TiO_3$ Ceramics by Bond Valence, *Jap. J. Appl. Phy.*, **43**, 219–22 (2004).

[16]L. Wu, D. Xiao, J. Wu, Y. Sun, D. Lin, J. Zhu, P. Yu, Y. Zhuang and Q. Wei, Good Temperature Stability of $K_{0.5}Na_{0.5}NbO_3$ Based Lead-free Ceramics and Their Applications in Buzzers, *J. Euro. Ceram. Soc.*, **28**, 2963-68 (2008).

[17]J. Tellier a, B. Malic, B. Dkhil, D. Jenko, J. Cilensek and M. Kosec, Crystal Structure and Phase Transitions of Sodium Potassium Niobate Perovskites, *Solid State Sciences*, **11**, 320-24 (2009).

OXYNITRIDES AS NEW FUNCTIONAL CERAMIC MATERIALS

Shinichi KIKKAWA*, Teruki MOTOHASHI and Yuji MASUBUCHI
Graduate School of Engineering, Hokkaido University,
N13W8 Kita-ku, Sapporo 060-8628, Japan
*kikkawa@eng.hokudai.ac.jp

ABSTRACT
SiAlON has been only one oxynitride as a useful ceramic material. It has been used as structural ceramics. Recently it has found out an additional application as phosphor materials in light emitting diode. We have developed a new preparation method through solution route for various kinds of metal oxynitrides. Magnetic and optical materials were found out in gallium oxynitride and its nanofiber was also obtained. New phosphor material could be developed in aluminum oxynitride. Structural investigation was performed on $EuTaO_2N$ perovskite as a model for potential lead-free dielectric $ATaO_2N$ (A=Sr, Ba). Possible new superconductor was also found out in niobium related oxynitride.

INTRODUCTION
Various kinds of metal oxides have been widely used as functional ceramic materials, especially spinel ferrites as magnetic materials, perovskite dielectrics as piezoelectric materials and so on. Limitted kinds of metal nitrides have found out their application in GaN semiconductor, Si_3N_4 structural ceramics and iron nitride magnetic materials. Information on nitrides is still very little and some papers are now appearing from a small research community in Europe and Japan[1-4]. Metal oxynitrides are in between oxides and nitrides. There should be a lot of variety in their combination. Their research as ceramics has not yet been developed probably because ceramic materials have been prepared by high temperature firing of oxide mixture. Nitrides have been obtained by successive nitridation after reduction of metal oxides. Low temperature nitridation of amorphous oxides through citrate route opened up a new field of oxynitrides.

The present paper reviews our recent developments on (1) preparation and doping of gallium oxynitride, (2) europium doped aluminum oxynitride phosphor, (3) europium tantalum oxynitride perovskite for dielectric material and (4) niobium-aluminum oxynitride superconductor prepared by ammonia nitridation of oxide precursor through citrate route.

PREPARATION AND DOPING OF GALLIUM OXYNITRIDE
Gallium nitride is a wide band gap semiconductor with Eg = 3.4eV. Other kinds of metal doping may give some additional functionality such as magnetism in magnetic semiconductor. Precursor obtained from gallium based citrate gel was nitrided in ammonia to study homogeneous doping for gallium oxynitride.

Citric acid was added to gallium nitrate aqueous solution and aged to obtain viscous gel. Oxide precursor obtained in its firing at 350°C in air was then nitrided in ammonia flow at 750°C. The product was isostructural to h-GaN. The diffraction lines were a little broad and its relative intensity was slightly different from h-GaN as shown in Fig. 1. Chemical analysis showed the product was represented as $(Ga_{0.89}\square_{0.11})(N_{0.66}O_{0.34})$[5]. Random distribution was expected for its gallium vacancy in the hexagonal crystal lattice from XRD pattern simulation as represented in Fig. 2.

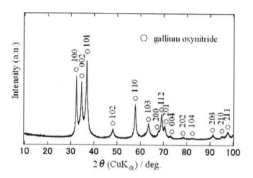

Similar preparation method was applied for other kinds of cationic doping. About 10 at% of Ga

Fig. 1 XRD pattern of gallium oxynitride prepared by gel-nitridation.

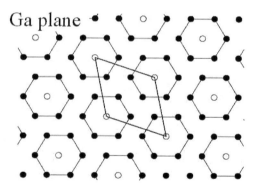

was substituted with Mn^{2+} and Li^+ which were normally in tetrahedral sites[5,6]. Solid solution limit was around 1 at% for Cr^{3+} and Eu^{3+} which prefer higher coordination number than six[7,8]. X-ray absorption spectrum suggested that manganese oxide cluster was formed in GaN matrix because of the thermal metastability in manganese nitrides[6]. (111) plane of bixbyte type Mn_2O_3 is epitaxial with c-plane of h-GaN. Moledular dynamic simulation suggested a possible cluster

Fig. 2 Gallium vacancy distribution in c-plane of wurtzite type gallium oxynitride.

formation of iron oxide in the iron doped gallium oxynitride[9]. Zinc could be doped to 33 at% into the gallium oxynitride lattice[10]. Lattice parameter changed linearly with the amount of zinc as shown in Fig. 3. Oxygen content also linearly increased with the amount of the doped Zn^{2+}. The optical band edge was independently at 3eV as shown in Fig. 4. Its sharpness was much improved by the Zn doping probably because the amount of gallium vacancy was much reduced by the codoping of Zn^{2+} and O^{2-} together into the GaN wurtzite lattice.

The gallium oxynitride became much more interesting when it was prepared in copresence of cobalt or nickel. A small amount of zinc blende type polymorph coexisted with the major wurtzite type gallium oxynitride[11]. Nano fibers depicted in Fig. 5 of the gallium oxynitride were grown as a mixture with its massive part.

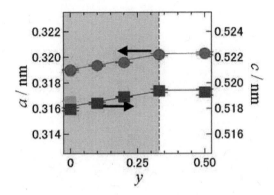

Fig. 3 Hexagonal lattice parameters of zinc dope gallium oxynitride. Green circle and grey square represent a and c literature values in JCPDS 50-1239.

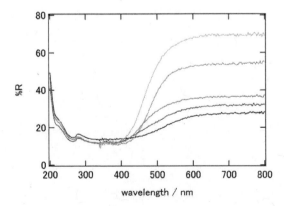

Fig. 4 Reflectance spectra of zinc doped gallium oxynitrides. Doping amounts were 0 (bottom), 0.1, 0.2, 0.33, 0.5 at% (top).

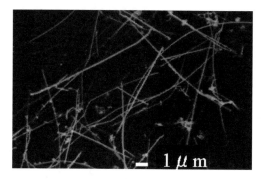

Fig. 5 Gallium oxynitride nano fibers grown in copresence of 3 at% Ni.

EUROPIUM DOPED ALUMINUM OXYNITRIDE PHOSPHOR

Spinel type AlON, $Al_{2.75}\square_{0.25}O_{3.74}N_{0.26}$, was obtained by ammonia nitridation of an oxide precursor prepared by peptizing a glycine gel with aluminum nitrate[12]. Post annealing at 1500℃ in flowing nitrogen was needed for its crystallization. The use of glycine instead of citric acid was important to obtain a white product without residual carbon. A similar preparation method was employed adding small amounts of europium below 10 at%. A strong blue emission was observed for products ranging from 0.5 to 3.0 at% Eu doping as shown in Fig. 6. The product with 0.5 at% doping had maximum emission intensity at 400nm for an excitation of 254nm. The products with 1 and 3 at% doping showed double maxima at 475nm and 520nm. These three emissions were due to the presence of divalent europium in the $EuAl_{12}O_{19}$ magnetoplumbite as an aluminum oxynitride impurity mixed with the AlON spinel major phase. The 1 at% Eu doped product exhibited expanded hexagonal lattice parameters (a= 0.5591nm and c= 2.236nm)

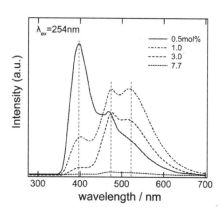

Fig. 6 Emission spectra for europium doped aluminum oxynitride products with various amounts of europium.

compared to the values for $EuAl_{12}O_{19}$ magnetoplumbite which was observed in the 7.7 at% doped product without any strong emission. The above spectrum change was related to the variable coordination around the interlayered Eu^{2+} with nitride and oxide ions in magnetoplumbite structure.

PHASE TRANSITION OF $Ca_{1-x}Eu_xTa(O,N)_3$ PEROVSKITE

Well crystallized product can be obtained in double metal oxynitrides as in the case of GaN-ZnO binary. Extra negative charge resulted in the nitride substitution with oxide can be compensated by the positive charge formed in the accompanied cation substitution. The simultaneous substitution has been reported on oxynitride perovskites; $(1-x)CaTaO_2N+xLaN\rightarrow$ $(Ca_{1-x}La_x)TaO_{2-x}N_x$ [13]. Large dielectric constants with slight temperature dependence have been reported on cubic $SrTaO_2N$ and $BaTaO_2N$ [14].

$CaTaO_2N$ crystallized in an orthorhombic system. The solid solution of $Ca_{1-x}Eu_xTa(O,N)_3$ could be prepared in a whole range of x [15]. It was orthorhombic in x<0.4 and changed to cubic in $0.4\leq x\leq1$. Their chemical analysis showed the oxygen and nitrogen contents were practically O_2N in the whole range. Europium was divalent in $EuTaO_2N$ in XANES spectrum. The cubic $EuTaO_2N$ transformed to tetragonal in its anneling at 1200℃ and returned back to cubic in its successive nitridation in ammonia. The nitrogen content was slightly less in the annealed sample. The cubic lattice in $EuTaO_2N$ might be an average of the tetragonal domains in small size as schematically shown in Fig. 7. This observation suggested that the cubic crystal lattices in $SrTaO_2N$ and $BaTaO_2N$ might be also formed with their small domains of crystallite in lower symmetry.

NIOBIUM-ALUMINUM OXYNITRIDE SUPERCONDUCTOR

Perovskite type crystal structure is formed in a combination of A-site cation in larger ionic size with B-site one in smaller size. Double cationic compounds with similar ionic size can be

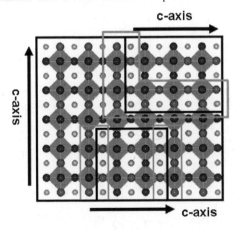

Fig. 7 Cubic $EuTaO_2N$ crystal lattice as the average of its tetragonal micro domains.

expected. Oxynitride preparation through citrate route was applied in niobium-aluminum binary system. Valency is variable between zero to five for niobium and fixed to trivalent for aluminum. A new oxynitride $(Nb_{0.56}Al_{0.44})(O_{0.38}N_{0.37}\square_{0.25})$ appeared in the system[16]. The product at 1000℃ showed an average structure in rock-salt type as refined in Table 1. It showed superconductivity with Tc=15K as shown in Fig. 8. Its volume fraction increased to 38% with its post annealing at 1100℃ in sealed evacuated tube. Superlattice related to the rock-salt structure appeared in the annealing.

Table.1 Refined structural parameters of $(Nb_{0.56}Al_{0.44})(O_{0.38}N_{0.37}\square_{0.25})$.

space group: *Fm-3m*   lattice parameter: $a = 0.43481(1)$ nm
$R_{wp} = 7.59\%$ $R_e = 3.46\%$ $S = 2.19$

Atom	Site	g	x	y	z	$B/10^{-2}$ nm^2
Nb	4a	0.557(5)	0	0	0	0.910(8)
Al	4a	0.443(5)	0	0	0	0.910(8)
O	4b	0.38	0.5	0.5	0.5	0.68(5)
N	4b	0.37	0.5	0.5	0.5	0.68(5)

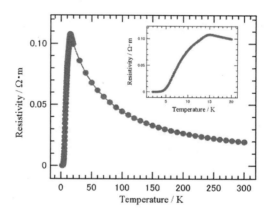

Fig. 8 Electrical resistivity of $(Nb_{0.56}Al_{0.44})(O_{0.38}N_{0.37}\square_{0.25})$.

CONCLUSIONS

Nitridation of amorphous oxide precursor through citrate route has opened up a new field of oxynitride research. Some examples of new compounds were reviewed in this manuscript. The preparation method will be applied not only for new compounds but also for the fabrications of thin films.

ACKNOWLEDGMENTS

This research was partly supported from the Grant-in-Aid for Scientific Research (A) and (B) of JSPS (grant nos. 21245047 and 19350098) and the Grant-in-Aid for Scientific Research on Priority Area from the Ministry of Education, Culture, Sports, Science and Technology (grant nos. 17042002 and 19018001).

REFERENCES

[1] R. Marchand, Y. Laurent, J. Guyader, P. L'Haridon, P. Verdier, *J. Eur. Ceram. Soc.*, **8**, 197-213 (1991).

[2] R. Marchand, in "Handbook on the Physics and Chemistry of Rare Earths" (K. A, Gschneider Jr. and L. Eyring Eds.), Elsevier Amsterdam, **25**, 51-99 (1998).

[3] R. Marchand, F. Tessier, A. L. Sauze and N. Diot, *Int. J. Inorg. Mater.*, **3**, 1143-46 (2001).

[4] F. Tessier, P. Maillard, F. Cheviré, K. Domen, S. Kikkawa, Optical properties of oxynitride powders, *J. Cer. Soc. Jpn.*, **117(1)**, 1-5 (2009).

[5] S. Kikkawa, K. Nagasaka, T. Takeda, M. Bailey, T. Sakurai and Y. Miyamoto, *J. Solid State Chem.*, **180**, 1984-89 (2007).

[6] S. Kikkawa, S. Ohtaki, T. Takeda, A. Yoshiasa, T. Sakurai and Y. Miyamoto, *J. Alloys & Compd.*, **450**, 152-56 (2008) .

[7] S. Yamamoto, S. Kikkawa, Y. Masubuchi, T. Takeda, H. Wolff, R. Dronskowski and A. Yoshiasa, *Solid State Commun.*, **147**, 41-45 (2008) .

[8] T. Takeda, N. Hatta and S. Kikkawa, *Chem. Lett.*, **35(9)**, 988-89 (2006).

[9] S. Yamamoto, S. Kikkawa, Y. Masubuchi, T. Takeda, M. Okube, A. Yoshiasa, M. Lumey and R. Dronskowski, *Mater. Res. Bull.*, **44**, 1656-59 (2009).

[10] A. Miyaake, Y. Masubuchi, T. Takeda and S. Kikkawa, *Mater. Res. Bull.*, **45**,505-08 (2010)

[11] A. Miyaake, Y. Masubuchi, T. Motohashi, T. Takeda and S. Kikkawa, *Dalton Trans.*, **39**,1-6(2010)

[12] S. Kikkawa, N. Hatta and T. Takeda, *J. Amer. Ceramic Soc.*, **91(3)**, 924-28 (2008).

[13] M. Jansen and H. P. Letschert, *Nature*, **404**, 980-82 (2000).

[14] Y. Kim, P. M. Woodward, K. Z. Baba-Kishi, and C. W. Tai, *Chem. Mater.*, **16(7)**, 1267-1276 (2004).

[15] Y. T. Motohashi, Y. Hamade, Y. Masubuchi, T. Takeda, K. Murai, A. Yoshiasa and S. Kikkawa, *Mater. Res. Bull.*, **44**, 1899–905 (2009).

[16] S. Yamamoto, Y. Ohashi, Y. Masubuchi, T. Takeda, T. Motohashi and S. Kikkawa, *J. Alloys and Compd.*, **482**, 160-63 (2009)

Microwave Materials

TERAHERTZ WAVE HARMONIZATION IN GEOMETRICALLY PATTERNED DIELECTRIC CERAMICS THROUGH SPATIALLY STRUCTURAL JOINING

Soshu Kirihara, Toshiki Niki1 and Masaru Kaneko
Joining and Welding Research Institute, Osaka University
11-1 Mihogaoka Ibaraki, 567-0047 Osaka, JAPAN

ABSTRACT

Two dimensional patterns composed of dielectric ceramics were fabricated in order to control terahertz (THz) waves effectively by using a micro-stereolithography. In this process, the photosensitive resin paste with titania particles dispersion was spread on a substrate with 10 μm in layer thickness by moving a knife edge, and two-dimensional images of UV ray were exposed by using digital micro-mirror device (DMD) with 2 μm in part accuracy. Through the layer by layer stacking process, the periodic structures composed of micro polygon tablets were successfully formed. The electromagnetic wave properties of these samples were measured by using a terahertz time domain spectroscopic (TDS) device. Forbidden band formations in transmission spectra and localization behaviors in the dielectric micro patterns were observed in the THz wave frequency range. And, the electric field profiles of localized modes were calculated by using transmission line modeling (TLM) of a finite difference time domain (FDTD) method.

INTRODUCTION

Photonic crystals with periodic variations in dielectric constants can exhibit forbidden gaps in electromagnetic wave spectra through Bragg diffraction [1-4]. The two dimensional photonic crystals with periodic arrangement of dielectric materials on plane substrates were well known to open the band gaps limitedly for the parallel directions to the plane shape structures. In our recent investigations, micro patterns with periodically arranged square tablets of above 30 in dielectric constant could exhibit the clear forbidden bands in the transmission spectra toward the perpendicular direction to the plane samples. The fabricated dielectric patterns were considered to totally reflect the terahertz wave at the wavelength comparable to the optical thickness. In this investigation, the two dimensional periodic patterns of hexagonal tablets composed of acrylic resins with nano-sized titania particles were fabricated by using micro-stereolithography to realize the wave diffractions and resonations in the terahertz frequencies. The stereolithography system of a computer-aided design and manufacturing (CAD/CAM) was newly developed in our research group to realize a spatially structural joining of ceramic components in micrometer orders [5-9]. Filtering effects of the electromagnetic waves for a perpendicular direction to the dielectric patterns were observed through time domain spectroscopic measurements. These micro geometric patterns of extremely thin devices with a high dielectric constant were designed to concentrate the electromagnetic energies effectively through a theoretical simulation. In near future industries, terahertz waves with micrometer order wavelengths will be expected to apply for various types of novel sensors which can detect gun powders, drugs, bacteria in foods, micro cracks in electric devices, cancer cells in human skin and other physical, chemical and living events [10-15].

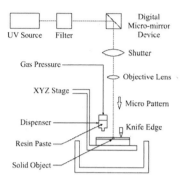
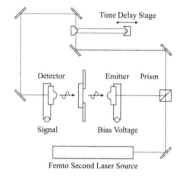

Figure 1   A schematically illustrated free forming system of a micro-stereolithography machine by using computer aided design and manufacturing (CAD/CAM) processes. (D-MEC Co. Ltd., Japan, SI-C 1000, http://www. d-mec.co.jp).

Figure 2   The schematically illustrated measuring system of a terahertz wave analyzer by using a time domain spectroscopic (TDS) detect method (Advanced Infrared Spectroscopy, Co. Ltd. Japan, J-Spec 2001, http://www. aispec.com).

EXPERIMENTAL PROCEDURE

The micro dielectric pattern was designed as the periodic structure with hexagonal tablets of 120 μm in edge length, 200 μm in thickness and 60 μm in interval. These micro tablets of 9×9=81 in numbers were arranged to form the extremely thin dielectric device of 2520×2520×200 μm in whole dimensions. The real sample was fabricated trough the micro-stereolithography system. A designed graphic model was converted for stereolithography (STL) data files and sliced into a series of two dimensional layers. These numerical data were transferred into the fabrication equipment (SIC-1000, D-MEC, Japan). Figure 1 shows the fabrication system. As the raw material, nanometer sized titania particles of 270 nm in average diameter were dispersed into a photo sensitive acrylic resin at 40 volume percent. The mixed slurry was squeezed on a working stage from a dispenser nozzle. This material paste was spread uniformly by a moving knife edge. Layer thickness was controlled to 5 μm. Ultra violet lay of 405 nm in wavelength was exposed on the resin surface according to the computer operation. Two dimensional solid patterns were obtained by a light induced photo polymerization. High resolutions in these micro patterns had been achieved by using a digital micro-mirror device (DMD). In this optical device, square aluminum mirrors of 14 μm in edge length were assembled with 1024×768 in numbers. Each micro mirror can be tilted independently, and cross sectional patterns were dynamically exposed through objective lenses as bitmap images of 2 μm in space resolution. After layer stacking and joining through photo solidifications, the periodical arrangements of the micro dielectric tablets were obtained. A bulk sample of the titania dispersed acrylic resin with the same material composition was also fabricated to measure the dielectric constant of the composite tablets. A terahertz wave attenuation of transmission amplitudes through the micro patterns were measured by using a terahertz time domain spectrometer (TDS) apparatus (Pulse-IRS, Aispec, Japan). Figure 2 shows the measurement system. Femto second laser beams were irradiated into a micro emission antenna patterned and formed on a semiconductor substrate to generate the terahertz wave pulses. The terahertz waves were transmitted trough the micro patterned samples perpendicularly. The dielectric constant of the bulk samples were measured through a phase shift counting. Diffraction and resonation behaviors in the dielectric pattern were calculated theoretically by using a transmission line modeling (TLM) simulator (Microstripes, Flomerics, EU) of a finite difference time domain (FDTD) method.

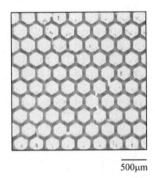

500μm

Figure 3   A micro periodic pattern with hexagonal tablets composed of titania dispersed acrylic resin.

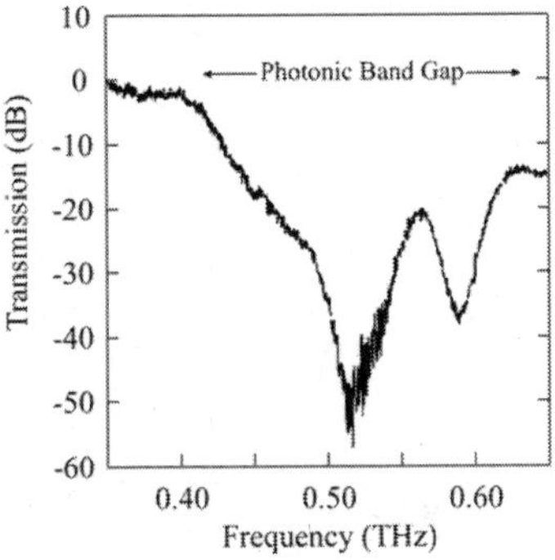

Figure 4   A transmission spectrum in the terahertz wave frequency through the dielectric micro pattern.

RESULTS AND DISCUSSION

The dielectric micro pattern with the periodic arrangement of the acrylic tablets with the titania particles dispersion was fabricated successfully by using the micro-stereolithography system as shown in Figure 3. Dimensional accuracies of the fabricated micro tablets and the air gaps were approximately 0.5 percent in length. The nanometer sized titania particles were verified to disperse uniformly in the acrylic resin matrix thorough a scanning electron microscope (SEM) observation. The dielectric constant of the composite material of the titania dispersed acrylic resin was measured as 40. Figure 4 shows transmission spectrum measured by using the TDS method. The measured result has good agreement with the calculated one by using the TLM method. Opaque regions were formed in both spectra form 0.40 to 0.65 THz approximately. Maximum attenuation was measured as about -50 dB in transmission amplitude. At the higher and lower frequency edges of this forbidden gap comparable to the optical thickness of the dielectric pattern, when the incident waves are reflected at the surface and bottom planes of the sample, total reflections are considered to be exhibited through the wave diffractions. As shown in Figure 5, two different standing waves vibrating in the air and the dielectric regions form the higher and the lower edges of the band gap.

The gap width can be controlled by varying geometric profile, filling ratio, and the dielectric constant of the tablets. As shown in Figure 5-(a), simulated electric field distributions indicate the characteristic energy concentrations in the micro pattern. The standing waves of the localized modes were formed in the air gaps between the hexagonal tablet arrangements.

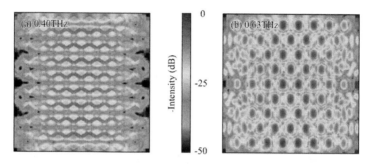

Figure 5   Cross sectional profiles of electric field intensities in the dielectric micro pattern calculated by using a transmission lime modeling method at higher and lower edge frequencies of the band gap.

CONCLUSIONS

Micro square tablets of acrylic resin with titania particles dispersions were arranged periodically in two dimensions by using a stereolithography system of a finely joining process. Fabricated micro pattern was verified to be able to exhibit a forbidden band of opaque region. A localized mode of a standing wave was clearly formed at the band edge frequencies to concentrate electromagnetic energies in the periodic arrangement of the dielectric constant. The terahertz waves are well known to resonate with various types of protein molecules, and expected to control the biological material syntheses by using frequency excitements through characteristic resonation effects. The fabricated micro pattern can include various types of solutions into their air gaps between the square tablets, therefore, it will be applied for novel micro reactors to create useful biological materials.

ACKNOWLEDGMENTS

This study was supported by Priority Assistance for the Formation of Worldwide Renowned Centers of Research - The Global COE Program (Project: Center of Excellence for Advanced Structural and Functional Materials Design) from the Ministry of Education, Culture, Sports, Science and Technology (MEXT), Japan.

REFERENCES

[1] Eli Yablonovitch: Inhabited Spontaneous Emission in Solid-state Physics and Electronics", Physical Review Letter, Vol. 58, No. 20, pp. 2059-2062, 1987.

[2] Sajeev John, "Strong Localization of Photons in Certain Disordered Dielectric Superlattices", Physical Review Letter, Vol. 58, No. 23, pp. 2486-2489, 1987.

[3] K. M. Ho, C. T. Chan, and C. M. Soukoulis, "Existence of a Photonic Gap in Periodic Dielectric Structures", Physical Review Letter", Vol. 65, No. 25, pp. 3152-3165, 1990.

[4] B. Temelkuran, Mehmet Bayindir, E. Ozbay, R. Biswas, M. M. Sigalas, G. Tuttle, and K. M. Ho, "Photonic Crystal-based Resonant Antenna with Very High Directivity", Journal of Applied Physics, Vol. 87, No. 1, pp. 603-605, 2000.

[5] Soshu Kirihara and Yoshinari Miyamoto: "Terahertz Wave Control Using Ceramic Photonic Crystals with Diamond Structure Including Plane Defects Fabricated by Micro-stereolithography", The International Journal of Applied Ceramic Technology, Vol. 6, No. 1, pp. 41-44, 2009.

[6] Weiwu Chen, Soshu Kirihara, and Yoshinari Miyamoto, "Fabrication of Three-Dimensional Micro Photonic Crystals of Resin-Incorporating TiO_2 Particles and their Terahertz Wave Properties", Journal of the American Ceramic Society, Vol. 90, No. 1, pp. 92-96, 2007.

[7] Weiwu Chen, Soshu Kirihara, Yoshinari Miyamoto, "Static Tuning Band Gaps of 3D Photonic Crystals in Subterahertz Frequencies", Applied Physics Letters, Vol. 92, pp. 183504-1-3, 2008.

[8] Hideaki Kanaoka, Soshu Kirihara, and Yoshinari Miyamoto "Terahertz Wave Properties of Alumina Microphotonic Crystals with a Diamond Structure", Journal of Materials Research Vol. 23, No. 4, pp. 1036-1041, 2008.

[9] Yoshinari Miyamoto, Hideaki Kanaoka, Soshu Kirihara "Terahertz Wave Localization at a Three-dimensional Ceramic Fractal Cavity in Photonic Crystals", Journal of Applied Physics, Vol. 103, pp. 103106-1-5, 2008.

[10] Martin Van Exter, Ch. Fattinger, and D. Grischkowsky: Terahertz Time-domain Spectroscopy of Water Vapor", Optics Letters Vol. 14, Iss. 20, pp. 1128-1130, 1989.

[11] Daniel Clery, "Brainstorming Their Way to an Imaging Revolution", Science, Vol. 297, pp. 761-763, 2002.

[12] Kodo Kawase, Yuichi Ogawa, Yuuki Watanabe, and Hiroyuki Inoue, "Non-destructive Terahertz Imaging of Illicit Drugs Using Spectral Fingerprints", Optics Express, Vol. 11, Iss. 20, pp. 2549-2554, 2003.

[13] R. M. Woodward, V. P. Wallace, D. D. Arnone, E. H. Linfield, and M. Pepper, "Terahertz Pulsed Imaging of Skin Cancer in the Time and Frequency Domain", Journal of Biological Physics, Vol. 29, No. 2-3, pp. 257-259, 2003.

[14] V. P. Wallace, A. J. Fitzgerald, S. Shankar, N. Flanagan, "Terahertz Pulsed Imaging of Basal Cell Carcinoma ex Vivo and in Vivo", The British Journal of Dermatology, Vol. 151, No. 2, pp. 424–432, 2004.

[15] Yutaka Oyama, Li Zhen, Tadao Tanabe, and Munehito Kagaya, "Sub-Terahertz Imaging of Defects in Building Blocks", NDT&E International, Vol. 42, No. 1, pp. 28-33, January, 2008.

TERAHERTZ WAVE PROPERTIES OF ALUMINA PHOTONIC CRYSTALS

Soshu Kirihara, Noritoshi Ohta, Toshiki Niki1 and Masaru Kaneko
Joining and Welding Research Institute, Osaka University
11-1 Mihogaoka Ibaraki, 567-0047 Osaka, JAPAN

ABSTRACT

Alumina photonic crystals with a diamond structure were fabricated to control terahertz waves by using micro stereolithography. Electromagnetic waves in a terahertz (THz) frequency range with micrometer order wavelength can be applied for various types of novel sensors. In the fabrication process, the photo sensitive resin paste with or without ceramic particles dispersion was spread on a grass substrate with 10 μm in layer thickness by using a mechanical knife edge, and two dimensional images of UV ray were exposed by using digital micro mirror device (DMD) with 2 μm in part accuracy. Through the layer by layer stacking, micrometer order three dimensional structures were formed. Dense Alumina structures were obtained by successive dewaxing and sintering in an air atmosphere. The electromagnetic wave properties of the photonic crystals were measured by THz wave time domain spectroscopy (TDS). The micrometer order periodic structures exhibited perfect band gaps in THz frequency range. The plane defects in the lattice structures formed highly localized mode in the band gaps. The propagation and localization behaviors of THz waves were simulated by transmission line modeling (TLM) methods.

INTRODUCTION

Photonic crystals composed of dielectric lattices form forbidden gaps in electromagnetic waves spectra [1-4]. These artificial crystals can totally reflect the light or electromagnetic waves at wavelengths comparable to the lattice spacings by Bragg deflection. The two different standing waves vibrating in the air and dielectric matrix form higher and lower frequency bands in the first and second Brillouin zones, respectively. The band gap widths can be controlled by varying the filling ratios, and dielectric constants of the lattices. Structural modifications by introducing defects can control the transmissions of electromagnetic waves [5-7]. The introduced structural defects in the periodic arrangements can localize the electromagnetic wave energies and form the transmission modes in the band gaps according to the sizes and dielectric constants of the defects.

Recently, we have newly developed micro stereolithography system to realize spatial joining of micrometer order ceramic structures with three dimensional distributions of dielectric materials. The final goal of our investigation is to control electromagnetic waves in a terahertz frequency range with micrometer wavelength effectively. In near future, the terahertz wave will be expected to apply to various types of novel sensors which can detect gun powders, drugs, bacteria in foods, micro cracks in electric devices, cancer cells in human skin and other physical, chemical and living events. To control terahertz waves effectively, micrometer sized electromagnetic devices composed of dielectric ceramics applying for cavities, filters and antennas will be necessary [8,9].

The photonic crystal with a diamond structure can form the perfect band gap which opens for all crystal directions [10,11]. We have fabricated the millimeter sized dielectric photonic crystal with a diamond structure to control the microwave by using the stereolithography of a structural joining process [12-14]. In our recent study, micrometer sized alumina lattices with a diamond structure were fabricated by using a newly developed micro stereolithography system [15-22]. These photonic crystals showed the perfect band gaps, which prohibited the terahertz wave propagation in all directions. Recently, we successfully fabricated a twinned diamond structure with a plane defect between the mirror symmetric lattice patterns. A localized mode to transmit the terahertz wave selectively was formed in the photonic band gap. In this study, transmission spectra of the terahertz wave through the diamond structures with the alumina lattices were measured. The selective transmission mode in the band gaps were observed for the twinned diamond structure and discussed in relation to the terahertz wave localization at the plane defect.

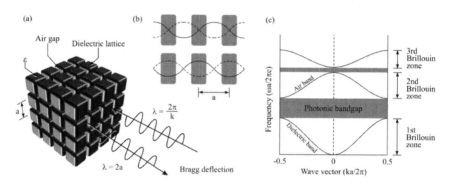

Figure 1   A schematic illustration of a band gap formation in an electromagnetic wave spectrum through a photonic crystal with periodic variations in dielectric constants by Bragg diffraction.

PRINCIPLE OF PHOTONIC CRYSTALS

The photonic crystals composed of dielectric lattices form band gaps for electromagnetic waves. These artificial crystals can totally reflect the light or microwave at wavelengths comparable to the lattice spacings by Bragg deflection as shown in Figure 1-(a). Two different standing waves (b) oscillating in the air and dielectric matrix form higher and lower frequency bands in the first and second Brillouin zones (c), respectively. The band gap width can be controlled by varying structure, filling ratio, and dielectric constant of the lattice. Structural modifications by introducing defects or varying the lattice spacing can control the propagations of the light or microwave.

The band diagram of the photonic crystal along symmetry lines in the Brillouin zone is drawn theoretically. Maxwell's equations (1) and (2) can be solved by means of plane wave expansion (PWE) method, where ω and c denote frequency and light velocity, respectively. Electronic and magnetic field $E\omega$ (r) and $H\omega$(r) are described with the following plane wave equations (2) and (4), respectively. The periodic arrangement of dielectric constant ε(r) can be obtained form the crystal structure. G and k are reciprocal vector and wave vector, respectively.

$$\left[\nabla \times \left(\frac{1}{\varepsilon(r)}\right) \nabla \times \right] H_{\omega}(r) = \left(\frac{\omega}{c}\right)^2 H_{\omega}(r) \cdots (1)$$

$$\left[\frac{1}{\varepsilon(r)} \nabla \times \nabla \times \right] E_{\omega}(r) = \left(\frac{\omega}{c}\right)^2 E_{\omega}(r) \cdots (2)$$

$$H_{k,n}(r) = \sum_{G} H_{k,n}(G) e^{i(k+G)\cdot r} \cdots (3)$$

$$E_{k,n}(r) = \sum_{G} E_{k,n}(G) e^{i(k+G)\cdot r} \cdots (4)$$

$$\frac{1}{\varepsilon(r)} = \sum_{G} \frac{1}{\varepsilon(G)} e^{iG\cdot r} \cdots (5)$$

Expected applications of the photonic crystal for the light and electromagnetic wave control in various wavelength ranges are shown in Figure 2. Air guides formed in the photonic crystal with nanometer order size will be used as the light wave circuit in the perfect reflective structure. When a light emitting diode is placed in an air cavity formed in a photonic crystal, an efficient laser emission can be enhanced due to the high coherent resonance in the micro cavity. While, millimeter order periodic structures can control microwaves effectively. Directional antennas and filters composed of photonic crystals can be applied to millimeter wave radar for intelligent traffic system and wireless communication system. The perfect reflection of millimeter wave by photonic crystal will be useful for barriers to prevent wave interference. The terahertz waves with micrometer order wavelength are expected to apply for novel sensors in medical or industrial inspection devices. The micro photonic crystals can applied for the terahertz wave cavities, filters and antennas.

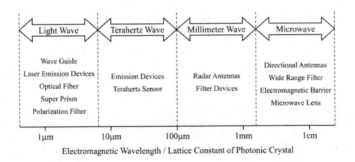

Figure 2   Expected applications of the photonic crystals from light wave to microwave band rages.

EXPERIMENTAL PROCEDURE

The three dimensional diamond lattice structures were designed by using a computer graphic application (Thinkdesign, Toyota Caelum, Japan). In the photonic crystal with diamond structure, the lattice constant was 1 mm. The whole structure was 6×6×2 mm³ in size, consisting of 6×6×2 unit cells. The aspect ratio of the lattices was designed to be 1.5. The graphic data was converted into the stereolithography format. After the slicing process of the graphic model into a series of two dimensional cross sectional patterns, this data was transferred to the micro stereolithography equipment (SI-C1000, D-MEC, Japan,). In our system, photo sensitive acrylic resins dispersed with alumina particles of 170nm in diameter at 40 volume % were fed over substrates from dispenser nozzles. The highly viscous resin paste was fed with controlled air pressure. It was spread uniformly by moving a knife edge. The thickness of each layer was controlled to 10 μm. A two dimensional pattern was formed by illuminating visible laser of 405 nm in wavelength on the resin surface. The high resolution has been achieved by using a digital micro mirror device (DMD) and an objective lens. Figure 3 shows a schematic of the micro stereolithography system. The DMD is an optical element assembled by micro mirrors of 14 μm in edge length. The tilting of each tiny mirror can be controlled according to the two dimensional cross sectional data transferred form a computer. The three dimensional structures were built by stacking these patterns layer by layer. In order to avoid deformation and cracking during dewaxing, careful investigation for the heat treatment processes were required. The precursors with diamond structures were heated at various temperatures from 100 to 600 °C while the heating rate was 1.0 °C /min. The dewaxing process was observed in respect to the weight and color changes. The alumina particles could be sintered at 1500 or 1000 °C, respectively. The heating rate was 8.0 °C/min. The density of the sintered sample was measured by the Archimedes method. The microstructures were observed by optical microscope and scanning electron microscopy (SEM). The transmittance and the phase shift of incident terahertz waves were measured by using terahertz time domain spectroscopy (TDS) device J-Spec2001, Aispec, Japan). Measured terahertz properties were compared with simulations by using a transmission line modeling (TLM) program (Micro-Stripes, Flomerics, EU).

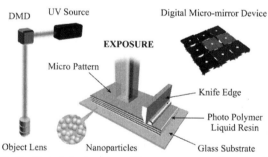

Figure 3   A schematically illustrated micro-stereolithography system of structural joining process.

RESULTS AND DISCUSSION

Three dimensional lattice structures composed of the alumina dispersed acrylic resin were processed exactly by using the micro stereolithography. The spatial resolution was approximately 0.5 %. Figure 2 shows a (100) plane of the sintered diamond structure composed of the micrometer order alumina lattice. The deformation and cracking were not observed. The linear shrinkage on the horizontal axis was 23.8 % and that on the vertical axis was 24.6 %. The relative density reached 97.5 %. Dense alumina microstructure was formed, and the average grain size was approximately 2 μm. The measured dielectric constant of the lattice was about 10. The terahertz wave attenuation of

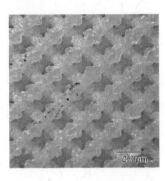

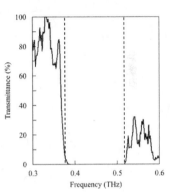

Figure 4 A micro dielectric photonic crystal composed of alumina lattices with diamond structure fabricated by stereolithography.

Figure 5 A measured transmission spectrum of terahertz waves for Γ-X <100> direction in the micro alumina photonic crystal.

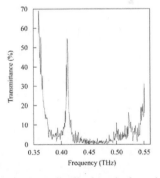

Figure 6 The twinned diamond photonic crystal with a plane defect between mirror symmetric lattices parallel to (100) layers.

Figure 7 A localized mode formed in the photonic band gap range through the twinned diamond structure of the alumina lattice.

the transmission amplitude through the alumina diamond structure for Γ-X <100> crystal direction is shown in Figure 3. The forbidden gap is formed at the frequency rage of from 0.37 to 0.52 THz. The dotted lines show the higher and lower band gap edges calculated by using the TLM method of a finite element method. The similar transmission spectra for Γ-K <110> and Γ-L <111> directions were obtained. These measured results of the photonic band gap frequencies were verified to have good agreements with the simulated ones within a tolerance of 5 %. A common band gap was observed in every crystal direction at the frequency range form 0.40 to 0.47 THz, where the electromagnetic wave cannot transmit through the lattices and is totally reflected in all directions. This common band was included in the calculated perfect photonic band gap by using the PWE method. In this theoretical calculation, the plane wave propagations were simulated for all directions in the periodic arrangements of the dielectric materials by solving Maxwell's equations as shown in Equations (1) to (5). And the photonic band distributions were drawn along the symmetry lines in the Brillouin zone. Figure 4 shows the twinned diamond structure composed of the mirror symmetric alumina lattices. The plane defect forms parallel to the (100) crystal layer. The transmission spectrum for the Γ-X <100> crystal direction of the twinned diamond structure is shown in Figure 5. The localized mode forms in the photonic band gap. At the peak frequency, the incident terahertz wave localized in the plane defect, and the amplified wave propagated to the opposite side. The three-dimensional photonic with the twinned diamond structure to form the band gap and the localized mode can be applied to the effective terahertz wave filters.

CONCLUSIONS

We have fabricated three dimensional micro photonic crystals with a diamond structure composed of alumina dispersed acrylic resin by using a micro stereolithography. By the careful optimization of process parameters regarding dewaxing and sintering, we have succeeded in fabricating dense alumina micro lattices. The sintered photonic crystals formed complete photonic band gaps at the terahertz region. A twinned diamond photonic crystal composed of alumina lattices with a plane defect between mirror symmetric lattice structures was also fabricated. A localized mode of a transmission peak was observed in the forbidden bands. In electromagnetic wave simulations, the localized mode with multiple reflections was formed in the plane defect between the twinned lattice patterns. These micro components of ceramic photonic crystals have potentials to be used as cavities, filters and antennas in a terahertz range.

ACKNOWLEDGMENTS

This study was supported by Priority Assistance for the Formation of Worldwide Renowned Centers of Research - The Global COE Program (Project: Center of Excellence for Advanced Structural and Functional Materials Design) from the Ministry of Education, Culture, Sports, Science and Technology (MEXT), Japan.

REFERENCES

[1] E. Yablonovitch, Phys. Rev. Lett. 58 (1987) 2059.

[2] K. Ohtaka, Phys. Rev. B, 19 (1979) 5057.

[3] S. John, J. Wang, Phys. Rev. Lett., 64 (1990) 2418.

[4] M. Soukoulis, Photonic Band Gap Materials, Kluwer Academic Publsher, Netherlands (1996).

[5] E. Brown, C. Parker, E. Yablonovich, J. Optical Society of America B, 10 (1993) 404.

[6] S. Noda, N. Yamamoto, H. Kobayashi, M. Okano, et. al, Appl. Phys. Lett., 75 (1999) 905.

[7] S. Kawakami, Photonic Crystals, CMC, Tokyo (2002).

[8] R. Woodward, V. Wallacel, D. Arnonel, E. Linfield, J. Biological Phys., 29 (2003) 257.

[9] M. Exter, C. Fattinger, D. Grischkowsky, Optics Lett., 14 (1989) 1128.

[10] K. H. Ho , C.T Chan, C. M. Soukoulis, Phys. Rev. Lett., 65 (1990) 3152.

[11] J. Haus, J. Modern Optics 41 (1994) 195.

[12] S. Kanehira, S. Kirihara, Y. Miyamoto, J. Am. Ceram. Soc., 88 (2005) 1461.

[13] S. Kirihara, Y. Miyamoto, K. Takenaga, M. Takeda, et. al., Solid State Comm., 121 (2002) 435.

[14] S. Kirihara, M. Takeda, K. Sakoda, Y. Miyamoto,2002 Solid State Comm., 124 (2002) 135.

[15] S. Kirihara, Y. Miyamoto, Int. J. Appl. Ceram. Tec., 6 (2009) 41.

[16] W. Chen, S. Kirihara, Y. Miyamoto, J. Am. Ceram. Soc., 90 (2007) 92.

[17] W. Chen, S. Kirihara, Y. Miyamoto, Appl. Phys. Lett., 92 (2008) 183504-1.

[18] H. Kanaoka, S. Kirihara, Y. Miyamoto, 23 (2008) 1036.

[19] Y. Miyamoto, H. Kanaoka, S. Kirihara, J. Appl. Phys. 103 (2008) 103106-1.

[20] S. Kirihara, T. Niki, M. Kaneko, J. Physics, in printing.

[21] S. Kirihara, T. Niki, M. Kaneko, Ferroelectrics, in printing.

[22] S. Kirihara, K. Tsutsumi, Y. Miyamoto, Science of Advanced Materials, in printing.

HIGH SYMMETRY BRINGS HIGH Q INSTEAD OF ORDERING IN Ba(Zn$_{1/3}$Nb$_{2/3}$)O$_3$: A HRTEM STUDY

Hitoshi Ohsato[1,2], Feridoon Azough[3], Eiichi Koga[4] and Isao Kagomiya[1] and Ken-ichi Kakimoto[1] and Robert Freer[3]
[1]Material Science and Engineering, Nagoya Institute of Technology, Gokiso-cho, Showa-ku, Nagoya, 466-8555, Japan.
[2]Dept. of Semiconductor and Display Engineering, BK21 Graduate School, Hoseo University, 165, Sechul-ri, Baebang-myeon, Asan-si 336-795, Chungnam, South Korea.
[3] Materials Science Centre, School of Materials, The University of Manchester, Grosvenor Street Manchester, Ml 7HS, UK
[4]Department of Engineering, Panasonic Electronic Devices Hokkaido Co. Ltd., Chitose 066-8502, Japan.

ABSTRACT
Ba(Zn$_{1/3}$Nb$_{2/3}$)O$_3$ (BZN), which exhibits a complex perovskite structure with an order-disorder transition at 1350 °C, has been examined in terms of ordering ratio and microstructure by the Rietveld method and by high resolution transmission electron microscopy (HRTEM), respectively. Two types of sample are used: sample A is a high temperature form sintered at 1400 °C for 100 h; sample B is a low temperature form, based on the high temperature form (sample A) which has been annealed at 1200 °C for 100 h. Ordering ratios of the two samples A and B are 27.6 and 54.2 %, respectively. By HRTEM observations, it was found that both samples exhibit disordered parts, without a superlattice, and ordered parts with a superlattice. This interpretation is supported by synchrotron XRPD data collected at Diamond in Oxford UK. Part of the ordering in sample A might have appeared during quenching. In the case of sample B, the ordering ratio was expected to be 80 % as shown for BZT in a previous paper.

INTRODUCTION

Recently, millimeter-wave dielectric ceramics have been explored for wide-ranging applications in fields such as wireless communication, ultrahigh speed wireless Local Area Network (LAN), Engineering Test Satellite (ETS) for high-speed mobile satellite communication and Pre-Crushed Safety (PCS) Systems on the Intelligent Transport Systems (ITS). The use of radio frequency (RF) for microwave communication is expanding to higher frequencies because of the shortage of frequency bands, and requests for high speed and high data transfer rate. As microwave dielectrics are expected to have a high quality factor Q (high Q) for such applications, we are clarifying the origin of high Q which is one of three important properties of the dielectrics[1,2]. We focus on the quality factor Q based on the crystal structure parameters such as ordering and symmetry[3,4]. Complex perovskite $A(B'_{1/3}B''_{2/3})O_3$ compounds with ordered B-site cations, due to long sintering times of more than 100 h, show high Q[5]. So, the ordering of cations has been believed to be the origin of high Q. We consider the origin of high Q may in fact be due to high symmetry, as seen in the following cases. In case 1: Qf values of the Ba(Zn$_{1/3}$Ta$_{2/3}$)O$_3$ (BZT) samples with different sintering time do not depend on the ordering ratios, which are saturated at about 80 %, though the value depends on the density and grain size[6]. In case 2: a composition deviating from a pure BZT composition shows higher Qf than that of pure BZT, though the ordering ratio is lower than that of pure BZT[7,8]. In case 3: Ba(Zn$_{1/3}$Nb$_{2/3}$)O$_3$ (BZN) which has an order-disorder transition at 1350 °C shows higher Qf on disorder ($Pm\overline{3}m$) above the transition temperature than below it ($P\overline{3}m1$), as shown in Fig. 1(a)[9]. Moreover, when a disordered sample sintered at 1400 °C for 100 h was annealed at 1200 °C for 100 h, the Qf did not change regardless of ordering, irrespective of grain size and density as shown in Fig. 1(b) and (c), respectively. In this paper, the ordering ratios and microstructure of BZN (in case 3) are defined using the Rietveld method and HRTEM, respectively[10].

EXPERIMENTAL

We examined BZN sintered at 1400 °C for 100 h (Sample A) and separately a sample of type A which was then annealed at 1200 °C for 100 h (Sample B); all were fabricated by the conventional method by Koga et al.[9] using high purity raw materials with 99.9 % or more purity, and vaporization was prevented[9]. The structures were identified by XRPD and the ordering ratios of both samples were confirmed by Rietveld method using RIETAN-2000[11] using XRPD patterns obtained by conventional JDX-8030 with CuKα radiation (measuring conditions: step width of 0.04 °, count time of 1 s per step and measuring range of 10 to 150 ° in 2θ). The samples were rechecked using the high resolution synchrotron radiation source, Diamond at Oxford UK. Microstructures of the samples were observed by scanning electron microscopy (SEM) and high resolution transmission electron microscopy (HRTEM, Field Emission Gun TECNAI G2). Specmen density was calculated from sample weight and dimension data. The grain size was determined from scanning electron micrographs of prepared surfaces using the linear intercept method. The dielectric properties were measured at 4 to 6 GHz using a network analyzer[12, 13]. The Q factor was determined by resonant cavity method in the TE$_{01\delta}$ mode.

RESULTS AND DISCUSSION

Samples of type A and B were first reported by Koga et al.,[9] We identified the present samples by conventional XRPD, as shown in Fig.2. As the superlattice lines are not clear, the high angle XRPD patterns around 2θ~115 ° are magnified, as shown in Fig. 2(b). The XRPD pattern of sample A (Fig. 2b) shows disorder because of the single peak of the 420 diffraction line. On the other hand, that of sample B shows order because of separation of the two peaks 422 and 226. The results are the same as in the previous study[9].

Fig.3 shows HRTEM results for sample A sintered at 1400°C for 100 h: (Fig. 3(a)) low magnification TEM image viewed along [110]$_{cubic}$ zone axis, (Fig. 3(b)) HRTEM image at the same orientation with Fast Fourier Transform (FFT) image of a disordered area without extra refrections along the [111]$_c$ direction. In Fig. 3(c) there is an HRTEM image with FFT of an ordered area; extra reflections exist at ($h \pm 1/3$, $k \pm 1/3$, $l \pm 1/3$) positions along [111]$_c$ direction. Both disordered and ordered areas exist in Fig. 3(d). If the sample contains a superlattice, additional reflections will appear at positions of ($h \pm 1/3$, $k \pm 1/3$, $l \pm 1/3$) as shown in Fig. 4(a), and 4(b). On the anti-phase domain boundary, both superlattices are superimposed to make four additional reciprocal lattice reflections as shown in Fig. 4(c). As the FFT in Fig. 3(b) shows no superlattice reflections, the area is disordered. On the other hand, as the FFT in Fig. 3(c) shows superlattice reflections, the area is ordered; some regions show both ordered and disordered areas (Fig. 3(d)). We could not confirm which type is predominant because of the small size of the sample (Fig. 3(a)). This compound sintered at 1400°C was expected to be disordered because the sintering temperature was above the order-disorder transition temperature. The ordering ratio was defined to be 27.6% by the Rietveld method. This result supports the existence of both ordered and disordered areas. As mentioned below, areas of both order and disorder coexisted under the transition temperature. It is therefore inferred that the sample texture is composed of ordered and disordered areas. Part of the ordering might have appeared during cooling (the cooling rate was 300 °C/h).

Fig.5 shows HRTEM results for sample B (same as sample A, but then annealed at 1200 °C for 100 h): Fig 5(a) is low magnification TEM image, Fig. 5(b) is an HRTEM image with FFT of ordered area, Fig. 5(c) is an ordered area showing an anti-phase domain, and Fig. 5(d) an area with both order and disorder. Thus the sample also contains both ordered and disordered areas (Fig. 5(b), 5(c) and 5(d)). Fig. 5(b) and 5(c) are from an ordered single domain and two ordered domains with anti-phase boundary, respectively. Importantly Fig. 5(d) is an HRTEM image of a grain with both disordered and ordered areas.

We collected high resolution XRPD patterns of sample A and B using synchrotron radiation as shown in Fig. 6. Superlattice diffraction 100t peaks (reciprocal lattice plane 100 in the trigonal crystal system) are observed on the both samples. The diffraction intensity for sample A is weaker than that of sample B. The intensities of the superlattice diffraction peaks are comparable with the ordering ratios: sample A and B 27.6 and 54.2 %, respectively, obtained by Rietveld method. Though the degree of ordering of sample B is high compared with that of sample A, it was expected to be about 80% ordering for a saturated sample, as the case of BZT.

We found that although the degree of ordering increased from 27.6 to 54.2% by annealing, the Qf values, the grain size and density were not affected (Fig. 1). As part of the disordered area of sample A (sintered above transitional temperature) changes to the low temperature form with ordering by annealing at a temperature below the transitional temperature, the Qf values are expected to increase by ordering. However, the Qf values only changed a little from 95,700 to 95,000 GHz. The effect of ordering is not enough to change the Qf value significantly.

CONCLUSIONS

Two samples of BZN with an order-disorder transition at 1350 °C were examined in terms of the degree of ordering and the microstructure: sample A was sintered at 1400 °C for 100 h;, sample B, was first treated as sample A then annealed at 1200 °C for 100 h, The Qf values, grain size and density of both samples A and B did not change significantly. By conventional XRPD, sample A was cubic and disordered; sample B was trigonal and partly ordered. By HRTEM, the samples A and B exhibited both ordered parts, with superlattice lines, and disordered parts, without superlattice lines. By the Rietveld method, the ordering ratios of the samples A and B were determined to be 27.6 and 54.2 %, respectively. Both samples A and B were examined by high resolution synchrotron XRPD, and showed evidence of superlattice diffraction peaks. The strength of the peak for the sample A was weaker than that of sample B, consistent with the ordering ratio. Part of the ordering of sample A may have developed during the quench procedure. On the other hand, the ordering ratio of sample B was low compared with BZT, which saturates at 80 %[6]. The reasons for this different behaviour need be clarified.

REFERENCES
[1]H. Ohsato, Microwave Materials with High and Low Dielectric Constant for Wireless Communications, *Mater. Res. Soc. Symp. Proc.*, **833**, 55-62 (2005).
[2]H. Ohsato, Research and Development of Microwave Dielectric Ceramics for Wireless Communications, *J. Ceram. Soc. Jpn.*, **113[11]**, 703-711 (2005).
[3]H. Ohsato, E. Koga, I. Kagomiya and K. Kakimoto, Origin of High Q for Microwave Complex Perovskite, *Key Eng. Mat.*, **421-422**, 77-80 (2010).
[4]H. Ohsato, E. Koga, I. Kagomiya and K. Kakimoto, Dense Composition with High Q on the Complex Perovskite Compounds, *Ferroelectrics*, **387**, 1-8 (2009).
[5]S. Kawashima, M. Nishida, I. Uchida and H. Ouchi, Ba(Zn$_{1/3}$Ta$_{2/3}$)O$_3$ Ceramics with Low Dielectric Loss at Microwave Frequencies, *J. Am. Ceram. Soc.*, **66**, 421-423 (1983).
[6]E. Koga, H. Moriwake, Effects of Superlattice Ordering and Ceramic Microstructure on the Microwave Q Factor of Complex Perovskite-Type Oxide Ba(Zn$_{1/3}$Ta$_{2/3}$)O$_3$, *J. Ceram. Soc. Jpn.*, **111**, 767-775 (2003) (Japanese).

[7] E. Koga, H. Moriwake, K. Kakimoto and H. Ohsato, Influence of Composition Deviation from Stoichiometric Ba(Zn$_{1/3}$Ta$_{2/3}$)O$_3$ on Superlattice Ordering and Microwave Quality Factor Q , *J. Ceram. Soc. Jpn.*, **113[2]**, 172-178 (2005) (Japanese).

[8] E. Koga, Y. Yamagishi, H. Moriwake, K. Kakimoto an H. Ohsato, Large Q Factor Variation within Dense, Highly Ordered Ba(Zn$_{1/3}$Ta$_{2/3}$)O$_3$ System, *J. Euro. Ceram. Soc.*, **26**, 1961-1964 (2006).

[9] E. Koga, Y. Yamagishi, H. Moriwake, K. Kakimoto an H. Ohsato, Order-Disorder Transition and its Effect on Microwave Quality Factor Q in Ba(Zn$_{1/3}$Nb$_{2/3}$)O$_3$ System, *J. Electroceram.*, **17**, 375-379 (2006).

[10] K-S. Hong, I-T. Kim and C-D. Kim, Order-Disorder Phase Formation in the Complex Perovskite Compounds Ba(Ni$_{1/3}$Nb$_{2/3}$)O$_3$ and Ba(Zn$_{1/3}$Nb$_{2/3}$)O$_3$, *J. Am. Ceram. Soc.*, **79**, 3218-3224 (1996).

[11] F. Izumi, and T. Ikeda, A Rietveld-Analysis Program RIETAN-98 and its Applications to Zeolites, *Mater. Sci. Forum*, **321-324**, 198-203(2000).

[12] B. W. Hakki and P. D. Coleman, A Dielectric Resonator Method of Measuring Inductive in the Millimeter Range, *IRE Trans. Microwave Theory & Tech.*, **MTT-33** , 402-410(1960).

[13] Y. Kobayashi and M. Kato, Microwave Measurement of Dielectric Properties of Low-loss Materials by the Dielectric Resonator Method, *IEEE Trans. Microwave Theory & Tech.*, **MTT-33**, 586-592 (1985).

[14] M. Terada, K. Kawamura, I. Kagomiya, K. Kakimoto and H. Ohsato, Effect of Ni Substitution on the Microwave Dielectric Properties of Cordierite, *J. Euro. Ceram. Soc.*, **27**, 3045-3048 (2007).

[15] H. Ohsato, I. Kagomiya, M. Terada and K. Kakimoto, Origin of Improvement of Q Based on High Symmetry Accompanying Si-Al Disordering in Cordierite Millimeterwave Ceramics, *J. Euro. Ceram. Soc.*, **30**, 315-318 (2010).

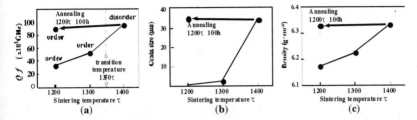

Fig. 1. Qf (a), grain size (b) and density (c) as a function of sintering temperature for BZN with transition temperature at 1350°C. Although the disordered sample with high Q was annealed at 1200°C for 100 h, the Qf was not improved.

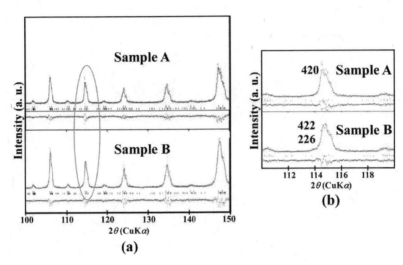

Fig. 2 XRPD patterns for BZN ceramics sintered at 1400 °C (sample A) and annealed at 1200 °C (sample B).

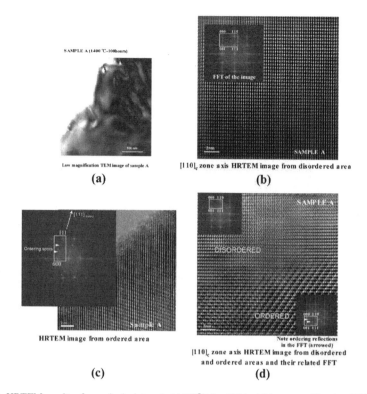

Fig.3 HRTEM results of sample A sintered at 1400°C for 100 h: (a) low magnification TEM image, (b) HRTEM images with FFT of disordered area, (c) ordered area, and (d) disordered and ordered areas .

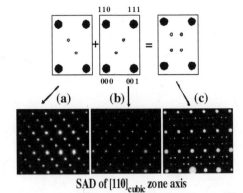

Fig. 4 Schematic representation and experimental electron diffraction patterns of ordering along [111] direction of the cubic unit cell: (a) ordering along two unique sets of (111) planes, (b) ordering along the other two sets of (111) planes, (c) is the superimposition of Fig. 4(a) and (b) at an antiphase domain boundary.

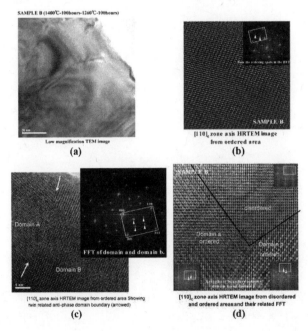

Fig.5 Results of HRTEM of sample B annealed at 1200 °C for 100 h: (a) low magnification TEM image, (b) HRTEM images with FFT of ordered area, (c), ordered area showing anti-phase domain, and (d) ordered and disordered areas.

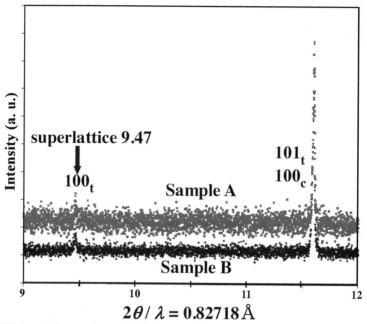

Fig. 6 High resolution synchrotron XRPD patterns (obtained with radiation of wavelength $\lambda = 0.82718$ Å) for sample A and B with superlattice peak 100t. Here, subscript t is trigonal, and c is cubic.

FLEXIBLE DESIGN OF COMPOSITE ELECTROMAGNETIC WAVE ABSORBER MADE OF ALUMINUM AND SENDUST PARTICLES DISPERSED IN POLYSTYRENE RESIN

Kenji Sakai, Yoichi Wada, Yuuki Sato, and Shinzo Yoshikado
Department of Electronics, Doshisha University
1-3 Tatara Miyakodani, Kyotanabe City, Kyoto 610-0321, Japan

ABSTRACT

For the purpose of designing electromagnetic wave absorbers with good absorption properties at frequencies above 1 GHz, the frequency dependences of the relative complex permeability μ_r^*, the relative complex permittivity ε_r^*, and the return loss were investigated for the composite made of both sendust (an alloy of Al 5%, Si 10%, and Fe 85%) and aluminum particles dispersed in polystyrene resin. The effects of the magnetic resonance caused by sendust and magnetic moments caused by sendust and aluminum on the frequency dependence of μ_r^* were observed in the composite made of both sendust and aluminum. Thus, it was found that the frequency dependence of μ_r^* for the composite made of sendust can be changed by adding aluminum particles in the composite made of sendust. Moreover, when each volume mixture ratio of sendust and aluminum was varied without changing the total volume mixture ratio of sendust and aluminum, the frequency dependence of μ_r^* was different despite the value of ε_r^* was almost the same. This result leads to a flexible design of an absorber with good absorption characteristics because the frequency dependence of μ_r^* can be controlled by selecting suitable volume mixture ratio of sendust and aluminum without changing the vale of ε_r^*. The composite made of both sendust and aluminum was found to exhibit a return loss of less than −20 dB in the frequency range of not only several GHz but also around 20 GHz. Furthermore, the reduction in weight of an absorber could be also possible since the aluminum is low mass density.

INTRODUCTION

The development of an electromagnetic wave absorber suitable for frequencies higher than 1 GHz is required with the increasing use of wireless telecommunication systems. To design a metal-backed single-layer absorber, the control of the frequency dependences of the relative complex permeability μ_r^* and the relative complex permittivity ε_r^* is important because the absorption of an electromagnetic wave is determined by both μ_r^* and ε_r^*. Thus, the frequency dependences of μ_r^*, ε_r^*, and the absorption characteristics for a composite made of magnetic material particles dispersed in an insulating matrix have been investigated[1-4]. In particular, μ_r', the real part of μ_r^* must be less than unity to satisfy the non-reflective condition of electromagnetic wave for a metal-backed single-layer absorber at frequencies above 10 GHz. We have reported that μ_r' for a composite made of soft magnetic material particles, such as sendust (an alloy of Al 5%, Si 10%, and Fe 85%) dispersed in polystyrene resin was less than unity and the absorption of large amount of electromagnetic waves

Table I. The volume mixture ratios of the composites made of sendust and aluminum particles.

	Sendust [vol%]	Aluminum [vol%]
Composite A	53	16.4
Composite B	25	25
Composite C	12.5	37.5

could be achieved at frequencies above 10 GHz[5]. The frequency dependence of μ_r^* for the composite made of soft magnetic material and polystyrene resin is similar to that required to satisfy the non-reflective condition. Therefore, the absorption with a wide frequency is expected if an optimum frequency dependence of μ_r^* can be obtained. However, the frequency dependence of μ_r^* is determined by the magnetic property, such as magnetic anisotropy. Thus, flexible control of the frequency dependence of μ_r^* is difficult using the composite made of a soft magnetic material and an insulating matrix. On the other hand, it has been also reported that the composite made of aluminum particles dispersed in polystyrene resin can control the values of both μ_r' and μ_r'' by the volume mixture ratio and the particle size of aluminum[6]. In addition, the value of μ_r' for the composite made of aluminum and polystyrene becomes less than unity. Therefore, the absorption of electromagnetic waves at any frequencies is possible. However, the values of μ_r' and μ_r'' for this composite are almost independent of frequency, hence the non-reflective condition is satisfied in a narrow frequency range and the bandwidth of absorption is narrow. From above results, it is considered that the frequency dependence of μ_r^* can be controlled flexibly so that the non-reflective condition is satisfied if both sendust and aluminum particles are dispersed in polystyrene resin. Furthermore, sendust and aluminum are low-cost materials and available in large quantities. Thus, the composite made of sendust and aluminum is suitable for practical use in an absorber and can avoid the problem of worldwide resource depletion.

In this study, the frequency dependences of μ_r^*, ε_r^*, and the absorption characteristics for a composite made of both sendust and aluminum particles dispersed in polystyrene resin were evaluated in the frequency range from 500 MHz to 40 GHz. Each volume mixture ratio of sendust and aluminum were varied fixing the total volume mixture ratio of sendust and aluminum to control the frequency dependence of μ_r^*.

EXPERIMENT

Commercially available sendust (Nippon Atomized Metal Powders Corporation, Al 5%, Si 10%, Fe 85%) particles and aluminum particles were used in this study. The average particle sizes (diameters) of the sendust were approximately 5 μm, and those of aluminum were approximately 8 μm. Chips of polystyrene resin were dissolved in acetone and the particles were mixed in until they were uniformly dispersed within the resin. The volume mixture ratios of sendust and aluminum particles are

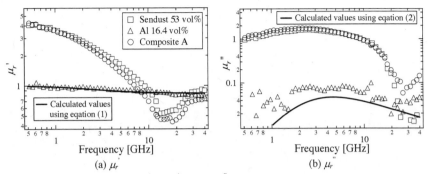

Figure 1. Frequency dependences of (a) $\mu_r{}'$ and (b) $\mu_r{}''$ for composites made of sendust or aluminum and composite A. Lines show values calculated using equations (1) and (2).

shown in table I. The mixture was then heated to melt the polystyrene resin and was hot-pressed at a pressure of 5 MPa to form a pellet. This was allowed to cool naturally to room temperature and was processed into a toroidal-core shape (outer diameter of approximately 7 mm, inner diameter of approximately 3 mm) for use in a 7 mm coaxial line in the frequency range 500 MHz to 12.4 GHz, or into a rectangular shape (P-band: 12.4-18 GHz, 15.80 mm × 7.90 mm, K-band: 18-26.5 GHz, 10.67 mm × 4.32 mm, R-band: 26.5-40 GHz, 7.11 mm × 3.56 mm) for use in a waveguide. The sample was mounted inside the coaxial line or waveguide using silver past to ensure that no gap existed between the sample and the walls of the line/waveguide. The complex scattering matrix elements $S_{11}{}^*$ (reflection coefficient) and $S_{21}{}^*$ (transmission coefficient) for the TEM mode (coaxial line) or TE_{10} mode (rectangular waveguide) were measured using a vector network analyzer (Agilent Technology, 8722ES) by the full-two-port method in the frequency range from 1 to 40 GHz. The values of $\mu_r{}^*$ ($\mu_r{}^* = \mu_r{}'-j\mu_r{}''$, $j=\sqrt{-1}$) and $\varepsilon_r{}^*$ ($\varepsilon_r{}^* = \varepsilon_r{}'-j\varepsilon_r{}''$) were calculated from the data of both $S_{11}{}^*$ and $S_{21}{}^*$. The complex reflection coefficient Γ for a metal-backed single-layer absorber was then determined from the values of $\mu_r{}^*$ and $\varepsilon_r{}^*$. The return loss R for each sample thickness was calculated from Γ using the relation $R = 20 \log_{10}|\Gamma|$. R was calculated at 0.1 mm intervals in the sample thickness range 0.1 to 30 mm.

RESULTS AND DISCUSSION
Frequency dependences of $\mu_r{}^*$ and $\varepsilon_r{}^*$ for the composites made of sendust or aluminum and both sendust and aluminum particles dispersed in polystyrene resin

Figure 1 shows the frequency dependences of $\mu_r{}'$ and $\mu_r{}''$ for composites made of sendust or aluminum and both sendust and aluminum (composite A). The values of $\mu_r{}'$ for the composite made of sendust decreased at frequencies above 500 MHz, reached a minimum around 14 GHz and increased with increasing frequency. Meanwhile, the values of $\mu_r{}''$ for the composite made of sendust reached a maximum around 2.5 GHz, decreased with increasing frequency and became almost zero at frequencies above 25 GHz. It is speculated that the frequency dependences of μ_r' and μ_r'' for the composite made of sendust is owing to the magnetic resonance of sendust, because sendust is magnetic material and similar frequency dependences of μ_r' and μ_r'' for the composite made of magnetic material has been observed[7].

The values of μ_r' for the composite made of aluminum decreased gradually with increasing frequency and the values of μ_r'' increased, reached maximum around 2.5 GHz and decreased gradually with increasing frequency. This phenomenon is owing to the magnetic moments generated by the eddy current flowing on the surface of aluminum particle[6]. In this case, the values of $1-\mu_r'$ and μ_r'' calculated for a electrical conductive metallic column of radius a and length $2a$ are derived in Appendix as

$$1-\mu_r'=\frac{\sigma\mu_0\omega^2 V\beta e^{-\frac{a}{\delta}}}{2a(1+\omega^2\beta^2)}\left[\delta(2\delta^2-2a\delta+a^2)(e^{\frac{a}{\delta}}-1)-a\delta(2\delta-a)\right] \tag{1}$$

$$\mu_r''=\mu_r'\frac{P}{\frac{\omega}{2}\mu_r'\mu_0\iiint H^2 dv}. \tag{2}$$

Here, H is the magnetic field in the column, V is the volume mixture ratio of the particles in the composite, P is the Joule loss caused by the eddy current loss per unit volume of the composite, δ is skin depth in aluminum (equation (A2)), σ is the electrical conductivity of aluminum, μ_0 is the permeability in free space, ω is the angular frequency of the electromagnetic wave. β, P, and the integral in the denominator of equation (2) is given by

$$\beta=\frac{\sigma\mu_0\delta}{a}\left(\frac{1}{2}a^2+\delta^2-\delta^2 e^{-\frac{a}{\delta}}-\delta a\right) \tag{3}$$

$$P=\frac{VA'^2 e^{-2\frac{a}{\delta}}}{\sigma a^2}\left[-\frac{\delta^2}{4}\left(e^{2\frac{a}{\delta}}-1\right)+\frac{a\delta}{2}e^{2\frac{a}{\delta}}\right]H_0^2 \tag{4}$$

$$\frac{\omega}{2}\mu_r'\mu_0\iiint H^2 dv=\left\{\frac{\omega}{2}\mu_r'\mu_0(1-V)+\frac{\omega\mu_r'\mu_0 V}{a^2}\left[\frac{1}{2}a^2+2A'\delta\left(\frac{1}{2}a^2-\delta a+\delta^2-\delta^2 e^{-\frac{a}{\delta}}\right)\right]\right.$$
$$\left.+\frac{\omega\mu_r'\mu_0 VA'^2\delta^2}{a^2}\left(\frac{1}{2}a^2-\frac{3}{2}\delta a+\frac{7}{4}\delta^2-2\delta^2 e^{-\frac{a}{\delta}}+\frac{1}{4}e^{-2\frac{a}{\delta}}\delta^2\right)\right\}H_0^2. \tag{5}$$

Here, H_0 is an incident magnetic field strength and $A'=-\dfrac{a\sigma\mu_0\omega^2\beta}{2(1+\omega^2\beta^2)}$.

The above qualitative results may also be applicable to spherical aluminum particles. It is found from equation (1) that $1-\mu_r'$ is proportional to V and depends on a and δ in the low frequency range. It is also found from equation (2) that μ_r'' depends not only on V and a but also on δ.

The lines shown in figures 1 (a) and (b) are the values calculated using equations (1) and (2). In the calculation of $\mu_r{'}$ and $\mu_r{''}$ using equations (1) and (2), a was modified by $\sqrt[3]{2/3}\, a$ assuming that the volumes of cylindrical and spherical particles are the same. It was found that the measured values of $\mu_r{'}$ and $\mu_r{''}$ for the composite made of aluminum qualitatively agreed with the calculated values. Therefore, the frequency dependences of $\mu_r{'}$ and $\mu_r{''}$ can be correctly determined using equations (1) and (2). Moreover, in the high frequency range, equations (1) and (2) are approximated as follows.

$$1 - \mu_r{'} = V \qquad\qquad (6)$$

$$\mu_r{''} = 2V\delta / a \qquad\qquad (7)$$

These results indicate that the value of $1 - \mu_r{'}$ is independent of the frequency and is proportional to V, and those of $\mu_r{''}$ increases proportionally to V and is inversely proportional to the square root of the frequency and a.

As shown in figure 1 (a), although the frequency dependence of $\mu_r{'}$ for composite A is similar to that for the composite made of sendust, the value of $\mu_r{'}$ for composite A became lower than that of the composite made of sendust at frequencies above 1.5 GHz. The frequency dependence of $\mu_r{''}$ for composite A was also similar to that for the composite made of sendust. However, the value of $\mu_r{''}$ for composite A was a little larger than that for the composite made of sendust at frequencies below 5 GHz, and increased with increasing frequency in the high frequency range above 25 GHz. It is speculated from above results that the frequency dependences of $\mu_r{'}$ and $\mu_r{''}$ for composite A is explained by the superposition of two effects; the natural magnetic resonance and the magnetic moment generated by the eddy current. Therefore, it was found that the frequency dependences of $\mu_r{'}$ and $\mu_r{''}$ for the composite made of sendust can be changed by adding aluminum particles in the composite. This result indicates that the control of frequency dependence of $\mu_r{*}$ is possible so that the non-reflective condition for a metal-backed single-layer absorber given by equation (8)[8] is satisfied to realize an absorber with a wide bandwidth.

$$\sqrt{\mu_r{*} / \varepsilon_r{*}}\, \tanh(\gamma_0 d \sqrt{\mu_r{*} \varepsilon_r{*}}) = 1 \qquad\qquad (8)$$

Here, γ_0 is the propagation constant in free space and d is thickness of the composite.

However, if the aluminum particles are added in the composite made of sendust particles dispersed in polystyrene resin, the total volume mixture ratio of sendust and aluminum particles increases and the increase in $\varepsilon_r{*}$ must be considered because the non-reflective condition is determined by not only $\mu_r{*}$ but also $\varepsilon_r{*}$. The frequency dependences of $\varepsilon_r{'}$ and $\varepsilon_r{''}$ for composites made of sendust, aluminum and composite A are shown in figure 2. The values of $\varepsilon_r{'}$ and $\varepsilon_r{''}$ for composite A were the largest of all composites, as shown in figure 2. This is because the volume mixture ratio of particles dispersed in polystyrene resin increased. Figure 3 shows surface optical micrographs of the composite made of sendust, aluminum and composite A. As shown in figure 3 (c), the distance between the particles in the composite decreased and the particles were partially in contact with each other. Therefore, the

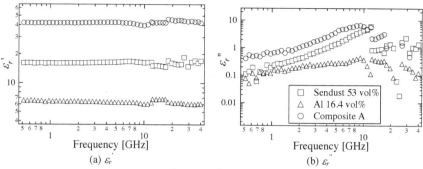

Figure 2. Frequency dependences of (a) $\varepsilon_r{'}$ and (b) $\varepsilon_r{''}$ for composites made of sendust or aluminum and composite A.

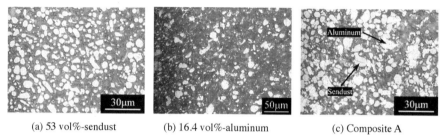

(a) 53 vol%-sendust   (b) 16.4 vol%-aluminum   (c) Composite A

Figure 3. Surface optical microphotographs of composites made of sendust or aluminum and composite A.

capacitance between the electrically conductive particles increased and the value of $\varepsilon_r{'}$ increased. In addition, the value of $\varepsilon_r{''}$ increases when the particles in the composite come in contact with each other since the average conductivity of the composite increases. Thus, composite A showed high values of both $\varepsilon_r{'}$ and $\varepsilon_r{''}$ because the total volume mixture ratio of sendust and aluminum increased.

Absorption characteristics of the composites made of both sendust and aluminum particles

The frequency dependence of the return loss R in free space was calculated from the measured values of $\mu_r{}^*$ and $\varepsilon_r{}^*$ for all samples. The absorber used for the calculation was a metal-backed single layer absorber and the incident electromagnetic wave was perpendicular to the surface. Figure 4 shows the frequency dependence of the return loss for composite A. The percentages shown in the graphs represent the normalized -20 dB bandwidth (the bandwidth Δf corresponding to a return loss of less than -20 dB divided by the absorption center frequency f_0). A value of -20 dB corresponds to the absorption of 99% of the electromagnetic wave power. The return loss of less than -20 dB at frequencies above 10 GHz could not be achieved for composite A although the return loss was less

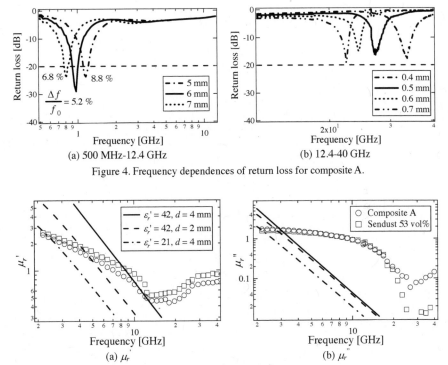

Figure 4. Frequency dependences of return loss for composite A.

Figure 5. Measured and calculated values of (a) $\mu_r{}'$ and (b) $\mu_r{}''$. Plots show the measured values for the composite made of 53 vol%-sendust and composite A. Lines show values calculated using equation (8).

than −20 dB around 1 GHz.

To examine the absorption of electromagnetic waves in the high frequency range, the values of $\mu_r{}'$ and $\mu_r{}''$ that satisfy the non-reflective condition given by equation (8) were calculated using the least squares method. The values of $\varepsilon_r{}'$ used for the calculation are independent of frequency and are the same as the measured value ($\varepsilon_r{}' = 42$), and the half of the measured value ($\varepsilon_r{}' = 21$). $\varepsilon_r{}''$ was assumed to be zero. The sample thicknesses used were 2 and 4 mm. Figure 5 shows the values of $\mu_r{}'$ and $\mu_r{}''$ calculated using equation (8), the measured values of $\mu_r{}'$ and $\mu_r{}''$ for composite A and those for the composite made of 53 vol%-sendust. The measured values of $\mu_r{}'$ for composite A agreed with the calculated line ($\varepsilon_r{}' = 42$ and $d = 4$ mm) in the frequency range from 9.5 to 13 GHz, as shown in figure 5 (a). Meanwhile, the measured values of $\mu_r{}''$ did not agree with the calculated values at this frequency range. Thus, the absorption of a large amount of electromagnetic wave power for composite A did not

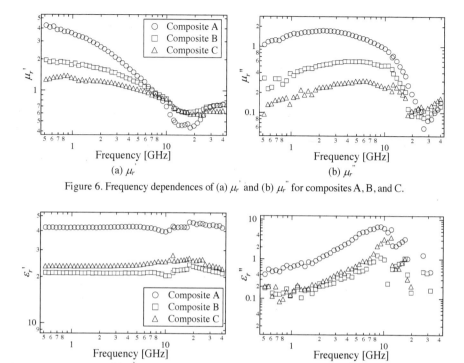

Figure 6. Frequency dependences of (a) $\mu_r{}'$ and (b) $\mu_r{}''$ for composites A, B, and C.

Figure 7. Frequency dependences of (a) $\varepsilon_r{}'$ and (b) $\varepsilon_r{}''$ for composites A, B, and C.

occur.

It is found from figure 5 that the slope of calculated lines of $\mu_r{}'$ and $\mu_r{}''$ using equation (8) is almost the same even if the values of $\varepsilon_r{}'$ and d are different. Therefore, the slope of the frequency dependences of $\mu_r{}'$ and $\mu_r{}''$ is important factor to realize an absorber with a wide bandwidth. As shown in figure 5 (a), an intersection occurred between the calculated and measured values of $\mu_r{}'$ for the composite made of sendust near 8.5 GHz. Meanwhile, the measured values of $\mu_r{}'$ for composite A agreed with the calculated line over a wide frequency range. This result was caused by the addition of aluminum particles in the composite made of sendust particles and the composite made of both sendust and aluminum enables the adjustment of the slope of $\mu_r{}'$ so that the non-reflective condition is satisfied in a wide frequency range. Moreover, in the high frequency range, equations (6) and (7) indicate that the value of $\mu_r{}''$ can be controlled by adjusting the particle size of aluminum without changing the value of $\mu_r{}'$. It was suggested that the composite made of both sendust and aluminum can realize an

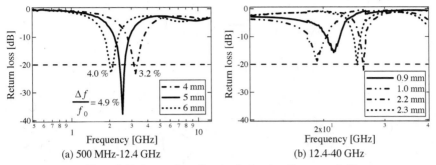

Figure 8. Frequency dependences of return loss for composite B.

absorber with a wide bandwidth in the high frequency range by selecting suitable volume mixture ratio and particle size of aluminum. In the next section, to control the frequency dependence of μ_r^* without an increase or decrease in ε_r^*, total amount of sendust and aluminum was fixed and each volume mixture ratio of sendust and aluminum was varied.

Frequency dependences of μ_r^* and ε_r^* for the composites made with various volume mixture ratios of sendust and aluminum

The frequency dependences of μ_r' and μ_r'' for composites A, B, and C are shown in figure 6. The values of μ_r' for composites B and C decreased with increasing frequency, reached a minimum and increased with increasing frequency. Meanwhile, the values of μ_r'' for composites B and C had a maximum in the frequency range of several GHz, decreased with increasing frequency and increased again at frequencies above 20 GHz. These frequency dependences are similar to those for composite A. Thus, the frequency dependence of μ_r^* for composites B and C is also explain by the superposition of the effect of natural magnetic resonance and magnetic moments by the eddy current. The values of μ_r' at frequencies below 6 GHz and μ_r'' at frequencies below 20 GHz were the largest for composite A and decreased in the order of composites B and C. The reason for this decrease is speculated to be that the amount of sendust in the composite decreased. On the other hand, at frequencies above 25 GHz, the values of μ_r' for composite C, which contains aluminum particles the most of three composites, were the smallest and those of μ_r'' for composite C were the largest. These results qualitatively agreed with the values calculated by equations (6) and (7). Thus, it is considered that the effect of magnetic moments by the eddy current is dominant in the high frequency range. Figure 7 shows the frequency dependences of ε_r' and ε_r'' for composites A, B, and C. The values of ε_r' and ε_r'' for composite B were almost the same as those for composite C. This is because the total amount of sendust and aluminum in composites B and C is the same. However, as shown in figure 6, the frequency dependences of μ_r' and μ_r'' for composite B were different from those for composite C. These results indicate that the

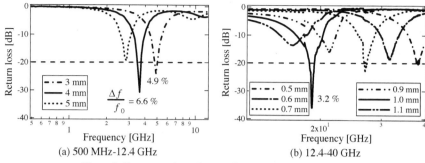

(a) 500 MHz-12.4 GHz (b) 12.4-40 GHz

Figure 9. Frequency dependences of return loss for composite C.

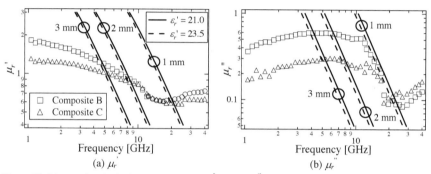

(a) $\mu_r{}'$ (b) $\mu_r{}''$

Figure 10. Measured and calculated values of (a) $\mu_r{}'$ and (b) $\mu_r{}''$. Plots show the measured values for the composites B and C. Lines show values calculated using equation (8).

frequency dependences of $\mu_r{}'$ and $\mu_r{}''$ can be controlled by adjusting each volume mixture ratio of sendust and aluminum without changing the values of $\varepsilon_r{}'$ and $\varepsilon_r{}''$. Thus, an absorber with good absorption characteristics can be designed easily using above results since the parameters which should be considered to satisfy equation (8) are only $\mu_r{}'$ and $\mu_r{}''$.

Absorption characteristics of the composites with various volume mixture ratios of sendust and aluminum

Figures 8 and 9 show the frequency dependences of the return loss for the composites B and C, respectively. Both composites B and C exhibited a return loss of less than −20 dB in the frequency range of several GHz. The normalized −20 dB band of these composites was between 3 and 7% and the sample thicknesses for which the return loss was less than −20 dB was several millimeter. These absorption characteristics can be also obtained for the composite made of soft magnetic materials dispersed in insulating matrix. For example, the return loss for the composite made of 53 vol%-sendust was less than -20 dB at a sample thickness of 5 mm, f_0 was 2 GHz, and the normalized -20 dB band was 13%. However, the mass density of the composite made of both sendust and aluminum was low because the mass density of aluminum is lower than that of sendust. The mass density of the composites B and C is approximately 2.4 and 2.1 g/cm^3, respectively, while that of the composite made of 53 vol%-sendust was approximately 3.5 g/cm^3. Therefore, the reduction in weight of absorber is possible using the composite made of both sendust and aluminum.

As shown in figure 9 (b), the return loss of composite C was also less than -20 dB around 20 GHz at a sample thickness of 1 mm. It was found that the composite made of both sendust and aluminum can be used as an absorber in the high frequency range. This is because the non-reflective condition is satisfied around 20 GHz, as shown in figure 10. Figure 10 shows the values of μ_r' and μ_r'' calculated by equation (8) and the measured values of μ_r' and μ_r'' for composite B and C. It is found from figure 10 that the calculated lines for $\varepsilon_r' = 21.0$ and $\varepsilon_r' = 23.5$ are the almost same when the sample thickness is the same. Thus, the difference in the values of ε_r' for composites B and C does not affect the values of μ_r' and μ_r'' that satisfy the non-reflective condition. Moreover, the slope of the frequency dependence of μ_r'' for composites B and C is similar to that of calculated lines in the frequency rage from 12 to 20. Therefore, if the values of μ_r' is controlled by changing the volume mixture ratio and the particle size of aluminum, the absorption of large amount of electromagnetic wave is expected.

CONCLUSION

The frequency dependences of μ_r' and μ_r'' could be changed by adding aluminum particles in the composite made of sendust and polystyrene resin. Therefore, an absorption with a wide frequency range is possible in the high frequency range. Moreover, when the total volume mixture ratio of sendust and aluminum was fixed and each volume mixture ratio of sendust and aluminum was varied, it was found that the different frequency dependences of μ_r' and μ_r'' can be obtained without changing the values of ε_r' and ε_r''. The composite made of 12.5 vol%-sendust and 37.5 vol%-aluminum exhibited a return loss of less than -20 dB in the frequency range of not only several GHz but also around 20 GHz. Furthermore, the mass density of the composite made of both sendust and aluminum was low and the reduction in weight of an absorber was proposed.

APPENDIX

Theoretical calculation of relative complex permeability

The values of μ_r' and μ_r'' are calculated theoretically as follows[9]. To simplify the discussion, the shape of an aluminum particle is approximated as a column of radius a and length $2a$, as shown in figure A (a). For an incident magnetic field strength of H_0 parallel to the central axis of the column, the eddy current density $J_\varphi(x)$ [A/m^2] may be defined as

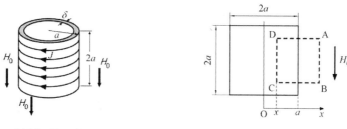

(a) Metallic column   (b) Sectional drawing

Figure A: Model of a electrical conductive metallic column used for calculation of μ_r^*.

$$J_\varphi(x) = Ae^{(-\frac{a-x}{\delta})}. \tag{A1}$$

Here, A is the proportional coefficient, x is the distance from the center of the column, and δ is skin depth in aluminum. δ is given by

$$\delta = \sqrt{\frac{2}{\omega\sigma\mu_0\mu_{Mr}'}}. \tag{A2}$$

Here, ω is the angular frequency of the electromagnetic wave, σ is the electrical conductivity of aluminum, μ_0 is the permeability in free space and μ_{Mr}' is the real part of the relative complex permeability of the aluminum (μ_{Mr}' is almost 1). It is assumed that a uniform magnetic field vector $\vec{H}'$ parallel to the central axis of the column is generated by $J_\varphi(x)$. When Ampere's circuital law is applied to the integral route of ABCD shown in figure A (b), the following equation is obtained, because $\vec{H}'$ is $\vec{0}$ along the route AB, $\vec{H}'\perp$ BC, and $\vec{H}'\perp$ AD.

$$\oint_{ABCD} \vec{H}'\cdot d\vec{s} = H'(x)\overline{DC} = \int_x^a J_\varphi(x)dx\,\overline{DC} \tag{A3}$$

Therefore, $H'(x)$ is obtained from equations (A1) and (A3).

$$H'(x) = \int_x^a Ae^{(-\frac{a-x}{\delta})}dx = Ae^{-\frac{a}{\delta}}\delta(e^{\frac{a}{\delta}} - e^{\frac{x}{\delta}}) \tag{A4}$$

For an electric field vector $\vec{E}$ and a magnetic flux density vector $\vec{B}$, Maxwell's equation and Stokes' theorem give the following equations.

$$\iint_S (\vec{\nabla} \times \vec{E}) \cdot d\vec{s} = -\frac{\partial}{\partial t}\iint_S \vec{B} \cdot d\vec{s} \tag{A5}$$

$$\iint_S (\vec{\nabla} \times \vec{E}) \cdot d\vec{s} = \oint_C \vec{E} \cdot d\vec{s} \tag{A6}$$

Here, C is a circle of radius x, whose center is O, and S is the area inside the integral route C. When the radius of C is a ($x = a$), the integral in the right side of equation (A5) is given by

$$\iint_S \vec{B} \cdot d\vec{S} = \iint_S (\vec{B}_0 + \vec{B}') \cdot d\vec{S} = \pi a^2 \mu_0 H_0 + \iint_S \vec{B}' \cdot d\vec{S} \tag{A7}$$

$$\iint_S \vec{B}' \cdot d\vec{S} = \int_0^a 2\pi x B'(x) dx$$

$$= 2\pi \mu_0 \int_0^a x H'(x) dx = 2\pi \mu_0 \int_0^a A e^{-\frac{a}{\delta}} \delta x (e^{\frac{a}{\delta}} - e^{\frac{x}{\delta}}) dx$$

$$= 2\pi \mu_0 A \delta \left(\frac{1}{2} a^2 + \delta^2 - \delta^2 e^{-\frac{a}{\delta}} - \delta a \right) \tag{A8}$$

Here, $\vec{B}_0$ is the external magnetic flux density vector and $\vec{B}'$ is the magnetic flux density vector generated by $J_\varphi(x)$. Also, the integral in the right side of equations (A6) is given by

$$\oint_C \vec{E} \cdot d\vec{s} = \oint_C E_\varphi(x) ds = 2\pi a E_\varphi(a) = \frac{2\pi a J_\varphi(a)}{\sigma} = \frac{2\pi a A}{\sigma} . \tag{A9}$$

Here, $E_\varphi(x)$ $(=J_\varphi(x)/\sigma)$ is the electric field at the circumference of the column. Thus, the following equation is obtained from equations (A7), (A8) and (A9).

$$\frac{2\pi a A}{\sigma} = -\frac{\partial}{\partial t} \left[\pi a^2 \mu_0 H_0 + 2\pi \mu_0 A \delta \left(\frac{1}{2} a^2 + \delta^2 - \delta^2 e^{-\frac{a}{\delta}} - \delta a \right) \right]$$

$$= -j\omega \left[\pi a^2 \mu_0 H_0 + 2\pi \mu_0 A \delta \left(\frac{1}{2} a^2 + \delta^2 - \delta^2 e^{-\frac{a}{\delta}} - \delta a \right) \right] \tag{A10}$$

A is obtained from equation (A10).

$$A = \frac{-j\omega a \mu_0 \sigma / 2}{1 + j\omega \mu_0 \delta \frac{\sigma}{a} \left(\frac{1}{2} a^2 + \delta^2 - \delta^2 e^{-\frac{a}{\delta}} - \delta a \right)} H_0 = \frac{-j\omega \alpha}{1 + j\omega \beta} H_0 \tag{A11}$$

Here, α and β are given by

$$\alpha = \frac{a \mu_0 \sigma}{2} \tag{A12}$$

$$\beta = \mu_0 \delta \frac{\sigma}{a} \left(\frac{1}{2} a^2 + \delta^2 - \delta^2 e^{-\frac{a}{\delta}} - \delta a \right) . \tag{A13}$$

When $a >> \delta$, β becomes

$$\beta = \mu_0 \delta \frac{\sigma}{a} \frac{1}{2} a^2 = \frac{\mu_0}{2} \delta \sigma a . \tag{A14}$$

Therefore, $J_\varphi(x)$ is obtained from equations (A1) and (A11).

$$J_\varphi(x) = \text{Re} \left(\frac{-j\omega \alpha}{1 + j\omega \beta} \right) e^{(\frac{a-x}{\delta})} H_0 = -\frac{\omega^2 \alpha \beta}{1 + \omega^2 \beta^2} e^{(\frac{a-x}{\delta})} H_0 = A' e^{(\frac{a-x}{\delta})} H_0 \tag{A15}$$

Here, A' is given by

$$A' = -\frac{\omega^2 \alpha \beta}{1 + \omega^2 \beta^2}.$$

(A16)

The magnetic moment m generated by $J_\varphi(x)$ is given by

$$m = 2a \int_0^a \pi x^2 J_\varphi(x) dx = -\frac{2\pi a \omega^2 \alpha \beta}{1 + \omega^2 \beta^2} e^{-\frac{a}{\delta}} H_0 \int_0^a x^2 e^{\frac{x}{\delta}} dx.$$

(A17)

If $2a$ is constant, the number N of cylindrical particles per unit volume of the composite is given by

$$N = V / 2\pi a^3.$$

(A18)

Here, V is the volume mixture ratio of the particles in the composite. If it is assumed that the direction of all magnetic moments is the same and that the eddy current loss is zero, the magnetization M is given by

$$M = Nm = -\frac{V\sigma\mu_0\omega^2\beta K_0 e^{-\frac{a}{\delta}}}{2a(1 + \omega^2\beta^2)} H_0.$$

(A19)

Here,

$$K_0 = \int_0^a x^2 e^{\frac{x}{\delta}} dx = \delta(2\delta^2 - 2a\delta + a^2)(e^{\frac{a}{\delta}} - 1) - a\delta(2\delta - a).$$

(A20)

Also, the following relation holds between the average magnetic flux density B and the magnetization M in the composite when M is assumed to be proportional to H_0.

$$M = B / \mu_0 - H_0 = (\mu_r^* - 1)H_0 = (\mu_r' - 1 - j\mu_r'')H_0$$

(A21)

Therefore, the following equation is obtained from equations (A19) and (A21).

$$1 - \mu_r' = \frac{\sigma\mu_0\omega^2 V\beta e^{-\frac{a}{\delta}}}{2a(1 + \omega^2\beta^2)} \left[\delta(2\delta^2 - 2a\delta + a^2)(e^{\frac{a}{\delta}} - 1) - a\delta(2\delta - a) \right]$$

(A22)

The Joule loss P, caused by the eddy current loss, per unit volume of the composite is

$$P = N\frac{1}{2}2a \int_0^a 2\pi x \frac{J_\varphi(x)^2}{\sigma} dx.$$

(A23)

Because $J_\varphi(x)$ is given by equation (A15), P is given by

$$P = \frac{VA'^2 e^{-2\frac{a}{\delta}}}{\sigma a^2} \left[-\frac{\delta^2}{4}\left(e^{2\frac{a}{\delta}} - 1\right) + \frac{a\delta}{2}e^{2\frac{a}{\delta}} \right] H_0^2.$$

(A24)

μ_r'' is defined as the ratio of the magnetic energy lost in one cycle and the magnetic energy accumulated in the column. Therefore, μ_r'' is given by

$$\mu_r'' = \mu_r' \frac{P}{\frac{\omega}{2}\mu_r'\mu_0 \iiint H^2 dv}.$$

(A25)

Here, H is the magnetic field in the column. The integral in the denominator of equation (A25) is obtained as follow.

$$\frac{\omega}{2}\mu_r'\mu_0\iiint H^2 dv = \frac{\omega}{2}\mu_r'\mu_0\iiint_{\substack{outside \\ cylinder}} H_0^2 dv + \frac{\omega}{2}\mu_r'\mu_0 N \iiint_{\substack{inside \\ cylinder}} \left(H_0 + H'(x)\right)^2 dv$$

$$= \frac{\omega}{2}\mu_r'\mu_0(1-V)H_0^2 + \frac{\omega}{2}\mu_r'\mu_0 N \int_0^a 2a\cdot 2\pi x\left(H_0^2 + 2H_0 H'(x) + H'(x)^2\right)$$

$$= \left\{\frac{\omega}{2}\mu_r'\mu_0(1-V) + \frac{\omega\mu_r'\mu_0 V}{a^2}\left[\frac{1}{2}a^2 + 2A'\delta\left(\frac{1}{2}a^2 - \delta a + \delta^2 - \delta^2 e^{-\frac{a}{\delta}}\right)\right]\right.$$

$$\left. + \frac{\omega\mu_r'\mu_0 V A'^2 \delta^2}{a^2}\left(\frac{1}{2}a^2 - \frac{3}{2}\delta a + \frac{7}{4}\delta^2 - 2\delta^2 e^{-\frac{a}{\delta}} + \frac{1}{4}e^{-\frac{2a}{\delta}}\delta^2\right)\right\}H_0^2 \qquad (A26)$$

ACKNOWLEDGEMENT

This work was supported by the Japan Society for the Promotion of Science (JSPS), the RCAST of Doshisha University, and Nippon Atomized Metal Powders Corporation.

REFERENCES

[1]S.-S. Kim, S.-T. Kim, Y.-C. Yoon, and K.-S. Lee, Magnetic, Dielectric, and Microwave Absorbing Properties of Iron Particles Dispersed in Rubber Matrix in Gigahertz Frequencies, *J. Appl. Phys.*, **97**, 10F905 (2005)

[2]A. L. Adenot-Engelvin, C. Dudeka, P. Toneguzzo, and O. Acher, Microwave Properties of Ferromagnetic Composites and Metamaterials, *J. Euro. Ceram. Soci.*, **27**, 1029-1033 (2007)

[3]S. M. Abbas, A. K. Dixit, R. Chatterjee, and T. C, Goel, Complex Permittivity, Complex Permeability and Microwave Absorption Properties of Ferrite-Polymer Composites, *J. Magn. Magn. Mater.*, **309**, 20-24 (2007)

[4]T. Kasagi, S. Suenaga, T. Tsutaoka, and K. Hatakeyama, High Frequency Permeability of Ferromagnetic Metal Composite Materials, *J. Magn. Magn. Mater.*, **310**, 2566-2568 (2007)

[5]K. Sakai, Y. Wada, and S. Yoshikado, Design of Composite Electromagnetic Wave Absorber Made of Soft Magnetic Materials Dispersed and Isolated in Polystyrene Resin, *Key Eng. Mater.*, **388**, 257-260 (2009)

[6]Y. Wada, N. Asano, K. Sakai, and S. Yoshikado, Preparation and Evaluation of Composite Electromagnetic Wave Absorbers Made of Fine Aluminum Particles Dispersed in Polystyrene Medium, *PIERS Online*, **4**, 838-845 (2008)

[7]T. Tsutaoka, Frequency Dispersion of Complex Permeability in Mn-Zn and Ni-Zn Spinel Ferrites and Their Composite Materials, *J. Magn. Magn. Mater.*, **93**, 2789-2796 (2003)

[8]Y. Naito and K. Suetake, Application of Ferrite to Electromagnetic Wave Absorber and Its Characteristics, *IEEE Trans. Microwave Theory Tech.*, **19**, 65-72 (1971)

[9]K. Sakai, Y. Wada, Y. Sato, and S. Yoshikado, Design of Composite Electromagnetic Wave Absorber Made of Fine Aluminum Particles Dispersed in Polystyrene Resin by Controlling Permeability, *PIERS Online*, in press

NEW PEROVSKITE OXIDES OF THE TYPE $(M_{1/4}Ln_{3/4})(Mg_{1/4}Ti_{3/4})O_3$ (M = Na, Li; Ln = La, Nd, Sm): CRYSTAL STRUCTURE AND MICROWAVE DIELECTRIC PROPERTIES

J.J. Bian[a], L.L. Yuan[a] and R. Ubic[b]
a:Department of Inorganic Materials, Shanghai University, China
b: Department of Materials Science and Engineering, Boise State University, U.S.A.

ABSTRACT

New complex perovskite compounds $La_{3/4}Na_{1/4}(Mg_{1/4}Ti_{3/4})O_3$(NLMT), $Nd_{3/4}Li_{1/4}(Mg_{1/4}Ti_{3/4})O_3$ (LNMT) and $Sm_{3/4}Li_{1/4}(Mg_{1/4}Ti_{3/4})O_3$ (LSMT) have been synthesized by solid state reaction process. Their crystal structure and microwave dielectric properties were investigated by using XRD Retiveld method, TEM, and network analyzer. All samples exhibited single perovskite phase with the same space group $Pbnm$ ($a^-a^-c^+$). No indication of cation ordering was found in these samples; however, the relative strength of superlattice reflections suggests that the degree of tilt of oxygen octahedra in LNMT and LSMT is higher than it is in NLMT, in agreement with Rietveld refinements of x-ray diffraction data. The dielectric permittivity decreased and $Q{\times}f$ value increased in the sequence of NLMT, LNMT and SLMT. All samples exhibited negative value of temperature coefficient of resonant frequency (τ_f).

INTRODUCTION

Compounds with perovskite structure have been widely studied not only because of interest in their crystal structural behavior, but also their wide variety of technologically important properties and applications, ranging from high-temperature superconductivity to ionic conduction and microwave dielectrics[1-3]. It is well known that several perovskite compounds have excellent microwave dielectric properties,[4-6] and they constitute a large share of the microwave dielectric materials commercially used as dielectric resonators in microwave communications. Since the perovskite structure can accommodate a variety of ions at both of the cation sites due to the high tolerance of the structural distortion, the formulations of the perovskite compounds are very diverse. Their dielectric properties can be tailored in a wide compositional range through multiple ion substitution on their A and /or B sites; hence, there has been a continuous effort to improve the dielectric properties of perovskite-based oxides or discover new dielectric materials by alloying at A and/or B sites. The removal of a requirement for an integer charge on the A/B-site leads to many new potential classes of perovskites. It is also possible to order the perovskite structure when mixed cations occupy the A and/or B-sites. It was generally believed that the ordered state is established primarily due to the charge difference between the cations but the difference in their radii is also important. High cation charge difference usually gives low dielectric loss for complex perovskites. In addition, the temperature coefficient of dielectric permittivity (TCε) for the perovskite compound is related to the tilting of the oxygen octahedra[7]. In particular, it was shown that a structure changed from an untilted cubic structure to that with anti-phase tilting of octahedra with a drastic change of both sign and value of TCε; therefore, the data on structure sequences in complex systems could be very useful in the development of new microwave dielectric ceramics.

The requirement of charge or size difference of the cations to stablize ordering of cations depends on the specific type of the ordering system. The stability of the cation ordering in perovskites is closely related to the specific changes in bulk chemistry.In this paper we report on the synthesis, crystal structure and microwave dielectric properties of new perovskite compounds with the formula $(M_{1/4}Ln_{3/4})(Mg_{1/4}Ti_{3/4})O_3$ (M = Na, Li; Ln = La, Nd, Sm). The potential possiblity of formation of 1:2

or 1:3 ordering in the system with an fractional charge on the A/B-site was also discusseed in this paper.

EXPERIMENTAL

$(Na_{1/4}La_{3/4})(Mg_{1/4}Ti_{3/4})O_3(NLMT)$, $(Li_{1/4}Nd_{3/4})(Mg_{1/4}Ti_{3/4})O_3(LNMT)$ and $(Li_{1/4}Sm_{3/4})(Mg_{1/4}Ti_{3/4})O_3(LSMT)$ ceramic samples were prepared by conventional solid-state reaction process from the starting materials including MgO (99.5%), Na_2CO_3(99.9%), Li_2CO_3(99.9%), TiO_2(99.7%), La_2O_3 (99.9%), Nd_2O_3 (99.9%) and Sm_2O_3(99.9%). The raw materials were weighed according to the above formalisms and milled with ZrO_2 balls in ethanol for 24h. The wet mixed powders were dried and calcined at the temperature of 1100-1300°C for 2h in an alumina crucible. The calcined powders were regrounded for 24h, dried, mixed with 7wt% PVA as binder and granulated. The granulated powders were uni-axially pressed into compacts 10mm in diameter and 4-5mm in height under the pressure of 100MPa. The compacts were sintered between 1300-1550 °C for 2h. In order to suppress the volatilization of lithium or sodium the compacts were muffled with powder of the same composition. The sintering conditions were optimized to get best densification and microwave dielectric properties. In order to study the effect of processing condition such as annealing on the cation ordering, some of the samples were annealed at a temperature 300-400°C lower than the sintering temperature for 10 hrs. Others were slow cooled after sintering (0.5°C/min).

Bulk densities of the sintered specimens were identified by the Archimedes' method. The phases were identified by using room temperature X-ray diffraction (XRD) with Ni-filtered CuKα radiation (40KV and 20mA, Model Dmax-RC, Japan). For Rietveld refinement of the x-ray patterns, the powder diffraction data were collected at room temperature with a step size $\Delta 2\theta = 0.02°$ over the angular range $15 \leq 2\theta(°) \leq 135$. The obtained data were refined by the Rietveld method using the Fullprof program.

All samples underwent thinning by conventional ceramographic techniques followed by ion milling (PIPS 691, Gatan, USA) for observation in the transmission electron microscope (JEOL 2100-HR with a LaB6 filament operating at 200kV). Microwave dielectric properties of the sintered samples were measured between 4 and 8GHz using network analyzer (Hewlett Packard, Model HP8720C, USA). The quality factor was measured by the transmission cavity method. The relative dielectric constant (ε_r) was measured according to the Hakki-Coleman method using the TE_{011} resonant mode, and the temperature coefficient of the resonator frequency (τ_f) was measured using invar cavity in the temperature range from 20 to 80 °C.

RESULTS AND DISCUSSIONS

Powder XRD patterns collected from the sintered samples are shown in Fig.1, which indicates that no secondary phases existed in these samples and that the compounds are consistent with perovskite structure. The splitting reflections suggest a noticeable distortion from the ideal cubic symmetry, and all compounds could be satisfactory indexed on the basis of an orthorhombic unit cell (*Pbnm*) (Fig.1).

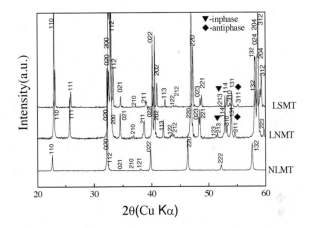

Fig.1 Powder XRD patterns of the NLMT, LNMT and LSMT sintered samples.

Retiveld's profile analysis method was applied for the refinement of NLMT, LNMT and LSMT x-ray diffraction results. A disordered structure model with the space group Pbnm was proposed and a pseudo-Voigt profile shape function was assumed. The background was described by interpolation of selected points. When refining the site occupancies, the unit cell content was restrained according to the nominal chemical composition. The observed, calculated and difference profiles for NLMT, LNMT and LSMT are plotted in Fig,2. Refined crystallographic results and main interatomic distances are given in Table 1 and Table 2, respectively. This model leads to the reliability factor R_{wp} of 12.6, 13.2 and 24.7 for NLMT , LNMT and LSMT, respectively and reasonable B-value for all atoms The larger R_{wp} factor for LSMT sample is mainly caused by the comparatively low integrated intensity of the peak in XRD raw data. The crystal structures of NLMT, LNMT and LSMT are shown in Fig.3. The possible tilt system could be $a^-a^-c^+$(Glazer's notation) according to Woodward's work on octahedal tilting in perovskites[8]. The superlattice reflections caused by the oxygen octahedral in-phase and anti-phase tilting are marked by solid triangle and diamond, respectively in Fig.1. The tilt angles along the [110] and [001] axis caculated according to the formulae[9]: $\theta=cos^{-1}(a/b)$ and $\psi = \cos^{-1}(\sqrt{2}a/c)$ are also listed in Table 1. Both θ and ψ inceased in the sequence of NLMT, LNMT and LSMT, which was predicted by the decrease of tolerance factor t (t-NLMT≈0.960, t-LNMT≈0.912 and t-LSMT=0.908). The interatomic distances given in Table 2 seem to indicate a little distortion of the oxygen octahedra.

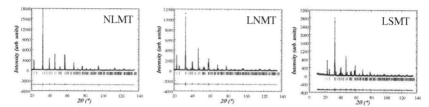

Fig.2 The observed, calculated and difference profiles for NLMT, LNMT and LSMT, respectively.

Table 1 Refined crystal data, fractional atomic coordinates, thermal parameters and occupancies of (a) NLMT (b)LNMT and (c) LSMT

(a)

Atom	Position	x/a	y/b	z/c	Biso (Å2)	Occupancy
Na	4c	0.0026	0.5004	0.2500	0.482	0.25
La	4c	0.0026	0.5004	0.2500	0.482	0.75
Mg	4a	0.00000	0.00000	0.00000	0.280	0.25
Ti	4a	0.0000	0.0000	0.0000	0.280	0.75
O1	4c	-0.0570	0.0410	0.2500	0.300	1
O2	8d	0.2450	0.2610	0.0322	0.160	2

a=5.52362Å, b=5.53918Å, c=7.80551Å, V=238.8199Å^3, R_p=10.9, R_{wp}=12.6, R_{exp}=10.4; Φ=4.30°, Ψ=1.93°

(b)

Atom	Position	x/a	y/b	z/c	Biso (Å2)	Occupancy
Li	4c	0.00795	0.03675	0.2500	0.243	0.2490
Nd	4c	0.00795	0.03675	0.2500	0.243	0.7498
Mg	4a	0.5000	0.0000	0.0000	0.590	0.25
Ti	4a	0.5000	0.0000	0.0000	0.590	0.75
O1	4c	0.92075	0.48668	0.2500	0.229	1
O2	8d	0.28814	0.28786	0.45628	0.148	2

a= 5.42789 Å, b= 5.50569 Å, c=7.7347Å, V=231.1459Å^3, R_p=11.4, R_{wp}=13.2, R_{exp}=11.25; Φ =9.64°, Ψ =6.57°

©

Atom	Position	x/a	y/b	z/c	Biso (Å2	Occupancy
Li	4c	0.9885	0.0467	0.2500	0.205	0.25
Sm	4c	0.9885	0.0467	0.2500	0.205	0.75
Mg	4a	0.0000	0.5000	0.0000	0.465	0.25
Ti	4a	0.0000	0.5000	0.0000	0.465	0.75
O1	4c	0.0898	0.4877	0.2500	0.0313	1
O2	8d	0.7066	0.2977	0.0440	0.0352	2

a= 5.38237 Å, b=5.51144Å, c=7.69430Å, V=228.2484Å^3, R_p=22.3, R_{wp}=24.7, R_{exp}=23.4, Φ=12.42 ° Ψ=8.40°

Table2 Bond distance in(Å)for NLMT, LNMT and LSMT

NLMT		LNMT		LSMT	
Na/La –O1	2.566(11)	Li/Nd-O1	3.070(6)	Li/Sm-O1	3.134(11)
Na/La –O1	3.013(11)	Li/Nd-O1	2.512(6)	Li/Sm-O1	2.486(11)
Na/La –O1	2.472(17)	Li/Nd-O1	2.358(8)	Li/Sm-O1	3.131(11)
Na/La –O1	3.071(17)	Li/Nd-O1	3.100(8)	Li/Sm-O1	2.293(11)
Na/La-O2	2.538(16) ×2	Li/Nd-O2	2.599(6) ×2	Li/Sm-O2	2.592(10) ×2
Na/La-O2	2.996(15) ×2	Li/Nd-O2	2.744(6) ×2	Li/Sm-O2	2.689(10) ×2
Na/La-O2	2.630(16)×2	Li/Nd-O2	2.379(6) ×2	Li/Sm-O2	2.347(10) ×2
Na/La-O2	2.910(15) ×2	Mg/Ti-O1	1.979(6) ×2	Mg/Ti-O1	1.985(3) ×2
Ti/Mg-O1	1.990(3) ×2	Mg/Ti-O2	1.981(6) ×2	Mg/Ti-O2	1.962(10) ×2
Ti/Mg-O2	1.996(19) ×2	Mg/Ti-O2	1.987(6) ×2	Mg/Ti-O2	2.011(10) ×2
Ti/Mg-O2	1.949(20) ×2				

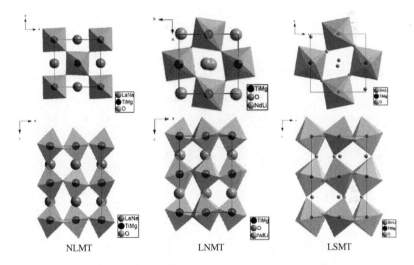

NLMT   LNMT   LSMT

Fig.3 Crystal structure of NLMT, LNMT and LSMT

The selected area diffraction patterns (SADPs) in Fig.4 were obtained from $(Li_{1/4}Nd_{3/4})(Mg_{1/4}Ti_{3/4})O_3$ (LNMT) samples. On the whole they agree with the refined structure in *Pbnm*. The [001]$_{pseudocubic}$ pattern shows evidence of in-phase tilting about the c axis but no evidence at all of antiparallel cation displacements along either a or b. This pattern also corresponds to [001] in the orthorhombic setting which is shown in reverse contrast. The [100]$_{pseudocubic}$ pattern (which is essentially identical to [010]) shows no evidence of in-phase tilts about a (or b); however, it does show evidence of antiparallel cation displacements parallel to c. This pattern corresponds to $[\bar{1}10]$ in the orthorhombic setting. The in-phase tilting story is summarized in the $[111]_{pseudocubic}$ pattern, which again shows evidence of in-phase tilts about only one axis. This pattern corresponds to the $[\bar{2}01]$ in the orthorhombic setting. The [110] pseudocubic pattern contains no evidence of antiphase tilting; however, the required α superlattice reflections are forbidden in $Pbnm$. Like the [100] pattern, this one also shows evidence of antiparallel cation displacements parallel to c. It corresponds to $[\bar{1}00]$ in the orthorhombic setting. The $[101]_{pseudocubic}$ pattern contains evidence of antiphase tilting about either a or c; however, as the $[001]_{pseudocubic}$ pattern has already shown us that octahedra are tilted in-phase about c, the antiphase tilts must be about b. This pattern corresponds to the $[\bar{1}11]$ in the orthorhombic setting. The $[011]_{pseudocubic}$ pattern is essentially identical to the [101] and confirms the existence of antiphase tilting about b. It seems certain that the structure of this material contains in-phase tilting about only one axis (the c axis in the $Pbnm$ setting) and antiphase tilts about the other two axes. This result is in agreement with the structures refined by the Retiveld method.

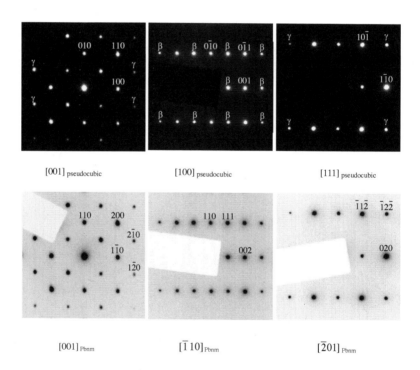

[001] pseudocubic   [100] pseudocubic   [111] pseudocubic

[001] Pbnm $[\bar{1}10]$ Pbnm $[\bar{2}01]$ Pbnm

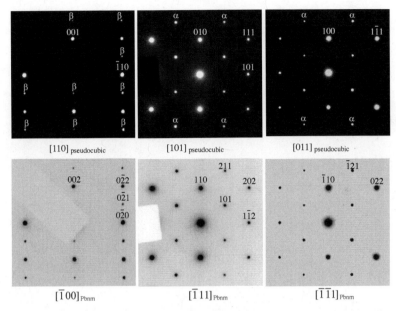

Fig.4 SADPs of LNMT sample. All indices are with respect to the pseudocubic cell.

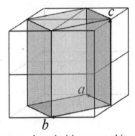

Fig.5 Geometrical relationship between the primitive perovskite unit cell and the orthorhombic one

The SADPs of the $(Na_{1/4}La_{3/4})(Mg_{1/4}Ti_{3/4})O_3$(NLMT) sample look similar to those of the LNMT samples in most respects (Fig.6). The [001] pseudocubic and [111] pseudocubic patterns show that the crystal structure also contains in-phase tilts about a single axis (the c axis in the *Pbnm* setting); however, the relative weakness of the γ superlattice reflections suggests that the degree of tilt is lower than it is in LNMT. This is consistent with the result of the above XRD analysis. The [101] pseudocubic pattern also contains evidence of antiphase tilting about the a axis. The space group *Pbnm* would suggest antiphase tilts about both a and b. Evidence of the common 90° twinning, which is quite widespread in the sample, is shown in Fig. 7. In addition, the grains appear very defective (Fig.7) in a way similar to the $Sm_{0.5}Li_{0.5}TiO_3$ grains previously observed[10]. As no indication of ordering was seen here (such ordering is forbidden in *Pbnm* anyway), and the same streaking in <100> directions is obvious in some diffraction patterns, these defects might also be 180° twin planes.

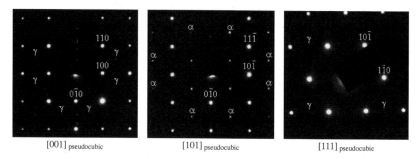

Fig.6 SADPs of NLMT sample. All indices are with respect to the pseudocubic cell.

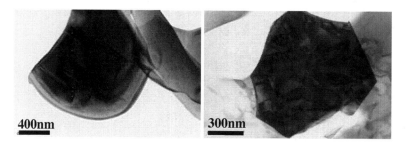

Fig.7 NLMT grain with defects on{100} and APBs.

The SADPs of the $(Li_{1/4}Sm_{3/4})(Mg_{1/4}Ti_{3/4})O_3$ (LSMT) sample are shown in Fig.8, and they look very similar to the others, especially LNMT, in terms of the degree of tilting. The two pseudocubic <100> patterns and the [111] pseudocubic show that the crystal structure contains in-phase tilts about a single axis (the *c* axis in the *Pbnm* setting). The [010] pseudocubic would correspond to $[\bar{1}\bar{1}0]$ in the *Pbnm* setting. As before, the [100] pseudocubic and [111] pseudocubic zones correspond to $[\bar{1}10]$ and $[\bar{2}01]$, respectively, in the *Pbnm* setting. The [101] pseudocubic pattern contains evidence of antiphase tilting about one or two axes as well. The space group *Pbnm* would require antiphase tilts of equal magnitude about two axes. The EDS data collected indicate that the cation composition is about Sm 48% (ideal is 43%), Mg 12% (ideal is 14%), and Ti 40% (ideal is 43%). These values are well within the range expected given the starting composition (Li is not detectable). No evidence of twinning or defects in LSMT samples could be found.

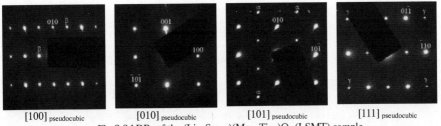

[100] pseudocubic   [010] pseudocubic   [101] pseudocubic   [111] pseudocubic

Fig.8 SADPs of the $(Li_{1/4}Sm_{3/4})(Mg_{1/4}Ti_{3/4})O_3$ (LSMT) sample

Both of the XRD and TEM analyses for the samples indicate no cation ordering occurred on A/B sites. Annealing or slow cooling processes conducted on NLMT samples had no effect on cation ordering at all. The ordering driving force was considered to be mainly charge and size difference between the cations, but cation stoichiometry is also important. It will produce two distinct A-site environments on the assumption that the B-site is 1:3 ordered[11]; however, in this case the ratio of the two A-sites (1:1) is incompatible with the cation stoichiometry (1:3 $Na^+/Li^+:La^{3+}/Nd^{3+}/Sm^{3+}$ or 1:2 $(Na_{3/4}La_{1/4})_{1/3}La_{2/3})$, and hence we can exclude the possiblity of 1:3 or 1:2 A-site ordering when the B-site is 1:3 ordered. Other possible 1:3 ordered site stoichiometry, e.g., $AMg_{1/4}Ti_{3/4}O_3$ (A-site disordering) was not observed either due to their relative small size difference (16%). This is consistent with the evidence of B-site disordering in $(La_{1/2}A^{2+}_{1/2})(Mg^{2+}_{1/4}Ti^{4+}_{3/4})O_3$ (A^{2+} = Ca, Sr, Ba)[11]. It was concluded that for a given charge difference, the formation of 1:3 order requires a larger size difference compared with the 1:2-ordered systems[11]. It is further supported by the following two examples: For a valence difference of 2, 1:1 order can be observed in $La(Mg_{1/2}Ti_{1/2})O_3$ but 1:3 order on the B-site can be stabilized only when the cations have a very large size difference like in $(La_{0.5}Ca_{0.5})(Ca_{1/4}Ti_{3/4})O_3$ (40%)[11]. For a large valence differnce of 4 such as in $Sr(Li_{1/4}Nb_{3/4})O_3$, the 1:3 ordering can be stabilized although size differnce(15.7%) is almost same as that in $AMg_{1/4}Ti_{3/4}O_3$[11]. In our case,the charge and size differences for all other possible 1:1 or 1:2 ordered site stoichiometry, e.g., $(Na_{1/2}La_{1/2})_{1/2}La_{1/2}$ $(Mg_{3/4}Ti_{1/4})_{1/3}$ $Ti_{2/3}O_3$, are even smaller than that in 1:3 ordered ones. Therefore disordering on A/B site is the only possible site stoichiometry for our samples.

Table 3   Microwave dielectric properties of sintered samples.

	Sintering temperature (°C)	ε_r	$Q \times f$(GHz)	τ_f(ppm/°C)	Relative Density(%)	Theoretical density(g/cm³)
NLMT	1500	47.8	4719	-31.4	94.1	5.57
LNMT	1300	43.2	5603	-117.6	98.5	5.61
LSMT	1300	38.5	12794	-79.7	96.1	5.97

The microwave dielectric properties of the sintered samples are listed in Table 3. Dielectric permittivity decreased and $Q \times f$ value increased in the sequence of NLMT, LNMT and LSMT. The decrease in dielectric permittivity is obviously due to the decrease in ionic polarizablity (α_{La}=6.07 ,

$\alpha_{Nd}=5.01$, $\alpha_{Sm}=4.74)^{12}$. The much lower $Q \times f$ value of NLMT (4719 GHz) compared with that of LSMT (12794 GHz) is mainly caused by the existence of widespread 90° twinning and heavily defective grains in NLMT samples, which could not be observed in the LSMT samples as discussed above. As shown in Table 3, temperature coefficients of resonant frequency, τ_f, of all samples are negative.

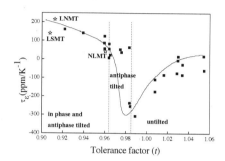

Fig.9 Relationship between the temperature coefficient of the permittivity TC_ε and the tolerance factor t obtained by Reaney et al. The data for the NLMT, LNMT and LSMT found in this work are also shown.

It is well known that the τ_f value for perovskite is mainly affected by the tilting of the oxygen octahedra. The tilting of oxygen octahedra is controlled by the tolerance factor of the perovskite structure. The relationship between temperature coefficients of dielectric constant (τ_ε) and tolerance factor (t) in Ba- and Sr- based complex perovskites was found by Reaney et al[7]. as shown in Fig.9. The data for the NLMT, LNMT and LSMT system found in this work are also appended in Fig.9. It can be seen that the tolerance factors of NLMT, LNMT and LSMT are all in the in-phase and anti-phase tilted region. All of the variation trends are in good agreement with those of Reaney et al.[7]. It has also been argued that the τ_f value increases linearly with the bond valence sum of oxygen if the nature of bond is similar in the same series[13]. The average bond valence sums of the oxygen in the above three compounds were calculated by using Fullprof-WinPLOTR. τ_f values of the sintered samples versus average bond valence sum of oxygen are shown in Fig.10. We can observe a good linear relation between τ_f value and bond valence. The τ_f value increased with increasing bond valence of oxygen.

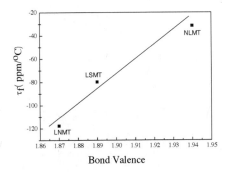

Fig.10 Variation of τ_f value as function of average bond valence sum of oxygen.

CONCLUSIONS

New complex perovskite series of NLMT, LNMT and LSMT have been successfully synthesized by solid state reaction process in this work. All samples exhibited single perovskite phase with the same space group, *Pbnm*. No indication of cation ordering was observed in these samples. The structures of these materials contained in-phase tilting about the *c* axis and antiphase tilts about two of the other axes. The tilt system could be $a^-a^-c^+$ (Glazer's notation), and the tilt angle increased in the sequence of NLMT, LNMT and SLMT. The dielectric permittivity decreased and $Q{\times}f$ value increased in the same sequence. All samples exhibited negative value of temperature coefficient of resonant frequency (τ_f). The τ_f value increased linearly with increasing bond valence of oxygen.

ACKNOWLEDGEMENTS

This work has been supported by the Natural Science Foundation of China (NSFC), (project number: 50572060), and part of the work (TEM) was also supported by the National Science Foundation of USA through the Major Research Instrumentation Program, Award Number 0521315, and the US Agency for International Development, award number PGA-P280420. The authors are also indebted to Steve Letourneau of Boise State University for TEM sample preparation."

REFERENCES

[1]S.A. Sunshine, D.W. Murphy, L.F. Schneemeyer, and J.V. Waszczak, The structure and properties of $Ba_2YCu_3O_6$, *Mater. Res. Bull.* **22**, 1007-13(1986).

[2]Y. Harada, Y. Hirakoso, H. Kawai, and J. Kuwano, Lithium ion conductivity of polycrystalline perovskite $La_{0.67-x}Li_{3x}TiO_3$ with ordered and disordered arrangements of the A-site ions, Solid State Ionics, **12**, 245-51(1999).

[3] H.Matsumoto, H.Tamura, and K.Wakino, Ba(Mg,Ta)O3–BaSnO3 high-Q dielectric resonator. *Jpn. J. Appl. Phys.*, **30**(9B), 2347–49(1991).

[4]K. Endo, K. Fujimoto, and K. Murakawa, Dielectric properties of ceramics in $Ba(Co_{1/3}Nb_{2/3})O_3$-$Ba(Zn_{1/3}Nb_{2/3})O_3$ solid Solutions, *J. Am. Ceram. Soc.*, **70**[9], C215-18(1987).

[5]I. M. Reaney, P. Wise, R. Ubic, J. Breeze, N. McN. Alford, D. Iddles, D.Cannel, and T. Price, On the Temperature Coefficient of Resonant Frequency in Microwave Dielectrics, *Philos. Mag.*, **81**, 501–10 (2001).

[6]H.Tamura, T.Konoike, Y.Sakabe, and K.Wakino, Improved high-Q dielectric resonator with complex perovskite structure, *J. Am. Ceram. Soc.*, **66**[4], C59-61(1984).

[7]E. L. Collar, I. M. Reaney, and N. Setter, Effect of Structural Changes in Complex Perovskites on the Temperature Coefficient of the Relative Permittivity, *J. Appl. Phys.*, **74**, 3414–25 (1993).

[8]P.M. Woodward, Octahedral tilting in perovskites. I. Geometrical considerations, *Acta Cryst.*, **B53**. 32-43(1997).

[9]Y. Zhao, D.J.Weidner, J.B.Parise, and D.E.Cox, Thermal expansion and structural distortion of perovskite-data for $NaMgF_3$ perovskite. Part II, *Physics of the Earth and Planetary Interiors*, **76**, 17-34(1993).

[10] J. J. Bian , K. Yan and R. Ubic, Structure and Microwave dielectric properties of $Sm_{1-x}Li_xTiO_3$, *Journal of Electroceramics* , **18**, 283-88 (2007).

[11]P.K. Davies,H.Wu, A.Y. Borisevich, I.E. Molodetsky, and L. Farber, Crystal chemistry of complex perovskites :New cation-ordered dielectric oxides, *Annu.Rev.Mater. Res.*, **38**, 369-401(2008).

[12]V.J. Fratello and C.D. Brandle, Calculation of dielectric polarizabilities of perovskite substrate materials for high-temperature superconductors, *J. Mater. Res.*, **9**[10], 2554-59(19940.

[13]L.L. Yuan and J.J. Bian, Structure and microwave dielectric properties of Lithium containing compounds with rock salt structure, *Ferroelectrics*, accepted.

UNDERSTANDING AND IMPROVING INSERTION LOSS AND INTERMODULATION IN MICROWAVE FERRITE DEVICES

David B. Cruickshank
TransTech Inc., a subsidiary of Skyworks Solutions Inc.
Adamstown, MD, USA

ABSTRACT

Intermodulation (IMD) and insertion loss in junction ferrite devices are not well understood from the point of view of the control of the properties of the ferrite material itself. It will be shown that the prime cause of IMD is insufficient local or general magnetic bias uniformity in the ferrite material, causing the material to show high IMD and insertion loss under certain conditions, and what steps can be taken to reduce both in practical devices.

INTRODUCTION

Ferrite circulators and isolators are used in communication systems over a wide frequency range, particularly from approximately 400MHz to 4GHz. Up to about 2.5 GHz, ferrites are typically biased above ferrimagnetic resonance with a range of choices of materials with narrow ferrimagnetic linewidths. Above 2.5GHz, because of magnet size and energy product limitations, below resonance operation is increasingly favored, and a different range of choices of materials are required (see Fig 1)

Device Mode	Recommended Magnetization in Gauss(G)/ Frequency (GHz)					
	0.4	1	2		3.1	4GHz
Above Resonance All power levels	YCaZrVIG 800G to 1850G or YAIIG 600G to 1780G					
Below Resonance, Low Power	YAIIG 150 to 550 Gauss		680G	800G	1000G	1200
Below Resonance, Medium Power	YGdAIIG 550 G		680G	800G	1000G	1200 G
Below Resonance, High Power	YGdHoAIG 550 G			800G	1000G	1200 G

Fig. 1 Materials for above and below operation from 400MHz to 4GHz.

ABOVE RESONANCE OPERATION

Above resonance it is evident that very narrow linewidths will be favored and this has produced a range of substituted yttrium iron garnet (YIG) compositions using V^{+5} on the tetrahedral site and a range of non-magnetic octahedral substitution that includes In^{+3}, Zr^{+4}, and Sn^{+4}. The effect of the latter substitutions is to reduce the linewidth from about 20 Oe in YIG to less than 10 Oe in very dense polycrystalline ceramics. However, the octahedral substitution also reduces the Curie temperature, typically to about 200C for these linewidths, whereas V^{+4} substitution does not.

Zr,V substituted YIG suffers from another potential disadvantage. If we look at the magnetization characteristics of YIG, it readily saturates with small values of applied field, giving it a very "square" hysteresis or BH loop, where squareness is defined as the ration of the remanence, Br, to the saturation magnetization, Bs when the applied field is approximately 10 times the coercive force, Hc. The degree of squareness manifests itself as a lack of magnetic saturation for materials with reduced squareness even under much higher applied fields such as those used in practical devices. For a fixed level of Zr, when increasing amounts of V are added to reduce the 4PiMs, there is a sharp drop in the squareness of the loop and corresponding difficulty in saturating the material (Fig2). Furthermore, external stress applied to the material can make it even less square, depending on the direction of the stress relative to the magnetization direction[1]. Note also that methods used in YIG substituted with Al^{+3} and $Gd^{+3,}$ such as substitution with Mn^{+3} to reduce magnetostriction, do not apply to Zr, V substituted YIG[1,2] because of the latter's near zero magnetocrystalline anisotropy.

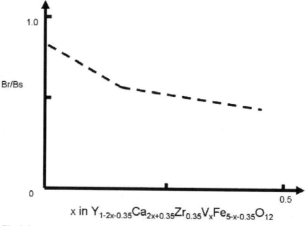

Fig.2 Squareness (Br/Bs) versus Vanadium content in Zr,V-doped YIG

Device Considerations

In a ferrite junction device the RF field is essentially symmetrical when biased. The operation of a typical device is shown in Fig 3. However the DC magnetic field is not uniform in a disk shaped ferrite because of its associated demagnetization field, where there is significant change in the demagnetization direction near the edge of the disk (Fig 4). The influence of ferrite shape on intermodulation has been reported in the literature by How[3] and Muira[4] and on insertion loss bandwidth by Schloemann[5] Uniformity of the field inside a ferrite part is only possible in a spherical or ellipsoidal shape, which is not practical in commercial devices. This is compounded by the behavior of the two incident RF signals which create intermodulation, as can be seen in Fig5. According to Wu[6], in a typical circulator, there is no intermodulation field interaction in the center of the disk, or in the disk on its top and bottom because of the stripline center conductor and ground plane. This leaves only the outside edge of the disk as the area where the 2 RF signals can interact. We thus have a situation where there are unfavorable demagnetization factors and interacting RF signals concentrated at the edge of the disk.

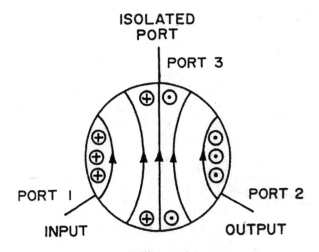

Fig.3 RF field distribution in a circulator.

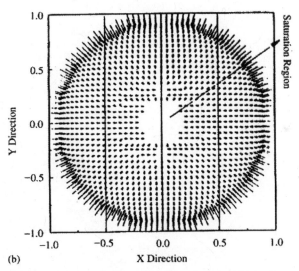

Fig.4 Demagnetization vectors in a thin magnetized ferrite disc.

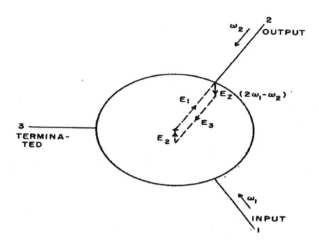

Fig.5 Intermodulation fields in a terminated circulator (isolator)

Experimentally, we can verify that this situation is the likely main cause of IMD by the following approaches using a typical isolator, seen in cross-section in Fig 6; The magnetic field is enhanced by a steel return path which is part of the device body, and the field is spread by pole pieces to improve the external field uniformity at the disc. The projected behavior of the field is shown by arrows. A space or a high dielectric constant ceramic ring is used to prevent excessive influence of the return path at the edge of the disk.

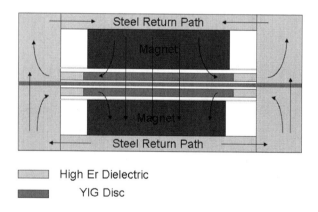

Fig.6 Magnetic circuit in a typical circulator

Experiments were carried out by
a) degrading the hysteresis loop locally 1) at the edge and 2) the top and bottom center of the disk, then compare the relative intermodulation. The hysteresis loop squareness, Br/Bs was degraded from >0.9 to <0.7 in both cases.
b) substituting different edged shapes of ferrite
c) substituting a ring of lower magnetization ferrite at the edge.
    a) was carried out by selectively sandblasting 1) the edge and 2) the top and bottom centers of the ferrite, in this case YIG (TT G113), see Figs 7 and 8. The BH loops were measured before and after until 1) and 2) were degraded equally in squareness. The loop degradation was found to be insensitive to the position of the area of degradation, only the total surface area, for the same degree of surface sandblasting.

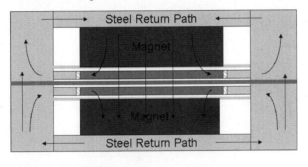

▨ High Er Dielectric
▨ Stressed Area

Fig 7 YIG discs stressed at edges.

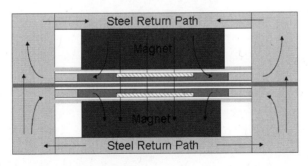

▨ High Er Dielectric
▨ Stressed Area

Fig 8 YIG discs stressed on top and bottom.

b) was arranged by preparing YIG disks with a 45 degree edge chamfer and comparing the disks with opposing chamfers (Fig 9) and chamfers in proximity to form an effective ellipsoid (Fig10)

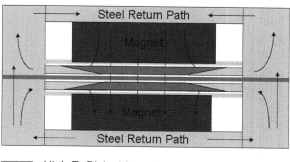

High Er Dielectric

Tapered YIG

Fig.9 YIG discs with opposing chamfers.

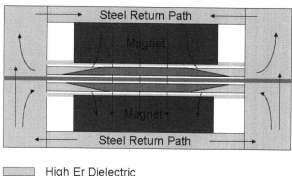

High Er Dielectric

Tapered YIG

Fig.10 YIG discs with chamfers in proximity.

c) was arranged by using a ring of a 1000 Gauss magnetization ferrite, YAlIG (TT G1010) replacing the outer part of the YIG (Fig 11).

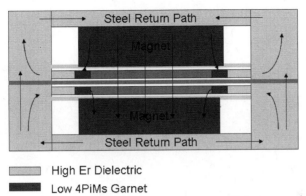

☐ High Er Dielectric

■ Low 4PiMs Garnet

Fig 11. Outer rings of YIG replaced by low 4PiMs rings

DISCUSSION

Above Resonance Results

The device with the disks sandblasted in the center showed virtually no degradation in IMD (< 5dB at 400MHz.) whereas the disk sandblasted at the edges showed a large relative degradation of more than 20dB, strongly suggesting the edge of the disk was by far the more sensitive area.

The disks with 45 degree chamfers showed quite surprising results. With chamfers where the top and bottom disks approximate the shape of an ellipsoid, the insertion loss increased sharply (>10dB) suggesting the edges of the disk were less magnetized and hence near ferromagnetic resonance because of effective demagnetization. The most likely cause of such behavior was the gap between the top and bottom disk preventing an effective ellipsoid forming, and the shape and lack of proximity of the DC magnetic field to the chamfer in that area of the device (Fig.10) By contrast, the opposing chamfers had normal insertion loss, suggesting the disks in that situation are more easily saturated by the closer proximity of the magnets (Fig 9).

When the outer part of the YIG was replaced by the ring of YAlIG, the results showed an improvement in IMD of more than 5 dB over the best results measured on any other configuration. This strongly suggested the outer ring's edge was more completely magnetically saturated because of its lower magnetization, keeping it well above resonance, and thus the interaction with the intermodulation field was much less.

Application to Higher Frequencies and the Below Resonance case.

Applying the same methodology at higher frequencies above resonance produces similar results but on a much reduced scale, typically <5dB at 900MHz when comparing degraded squareness with non-degraded squareness YIG. Although this could be due to differences in device geometry, applied field and so on, another possibility is less saturation at 400MHz, even though the material is biased above resonance in the lower frequency. Hence we cannot assume that above resonance operation guarantees saturation as the frequency is reduced, suggesting that the 4PiMs should be scaled to frequency at the same relative magnetic bias point to avoid IMD and potential insertion loss increases.

In the below resonance case it is well known[7] that insertion losses are higher due to "low field" losses unless the product of gamma, the gyromagnetic ratio times the magnetization is less than the lowest operating frequency when the shape and hence demagnetization factor are suitably taken into account. For YIG with typical device geometries this means that operation below about 5GHz in the below resonance condition will cause "low field" losses in the device[7]. By substituting materials with lower magnetizations, typically YAlIG's with 4PiMs of as low as 1200 Gauss, low field losses can be reduced or eliminated. IMD can be dealt with in the same way, although to ensure saturation it is useful to reduce the maximum 4PiMs further (Fig 12).

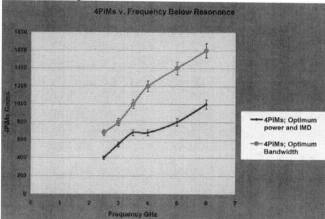

Fig.12 Selection of 4PiMs against frequency for IMD and bandwidth, below resonance.

CONCLUSIONS

In terms of selection of materials for ferrite devices which require very low insertion loss and IMD characteristics, there are at least two possible conclusions. One is that attention is required to the choice of saturation characteristics of materials used in situations where they might be sufficiently demagnetized to drop back into resonance or even also low field loss in the above resonance case, and low field loss in the below resonance case, because of the demagnetization factors operating at or near the edge of the disc. This would tend to favor materials with higher BH loop squareness if they have comparable ferromagnetic resonance linewidths, suggesting YIG and YAlIG materials might be preferable to YCaZrVIG, particularly since they also have higher Curie temperatures for the same magnetization. For this reason, process research is concentrated on further reducing the linewidths of YIG and YALIG to near the theoretical limit of about 15 Oe, determined by the magnetocrystalline anisotropy and magnetization[8]. Scaling the 4PiMs to frequency by Al substitution in YIG to avoid excessive IMD and insertion loss in the UHF region is also indicated.

The second conclusion is that in the above resonance case at least, IMD can be further reduced by replacing the outer part of the ferrite with a ring of lower magnetization material with otherwise similar characteristics. This is true even if a ring of high dielectric constant material is also used around the ferrite to reduce the proximity of the ferrite to the DC magnetic return path.

We can also speculate that all factors which degrade the BH loop properties of ferrite, whether they be compositional, microstructural, mechanical stress or external pressure induced, are the prime causes of intermodulation degradation when they influence the saturation of a ferrite device particularly at its edges. An approximate guide to that behavior for the above resonance device studied here is shown in Fig.13. To extrapolate to other frequencies, it would be necessary to scale the 4PiMs and bias point relative to resonance.

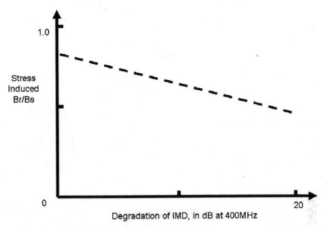

Fig.13 Approximate reduction in sqareness Br/Bs, all causes, versus IMD degradation.

ACKNOWLEDGEMENTS
 The author is grateful to Nikolai Volobouev of SDP components Inc. for many useful discussions and for microwave device testing, and to many staff members at TransTech for the materials' manufacture and testing, particularly Robert Burke and Rick Hahn.

REFERENCES
[1] G.F. Dionne, Effect of External Stress on Remanence Ratio and Anisotropy Field of Magnetic Materials, *IEEE Trans. Mag*.**5**, No 3, 596-600(1969)

[2] G.F. Dionne, Temperature and Stress Sensitivities of Microwave Ferrites, *IEEE Trans. Mag*. **8** No 3, 439-443 (1972)
[3] H.How, C.Vittoria, R.Schmidt, Nonlinear Coupling in Ferrite Circulator Junctions *IEEE MTT* **45** No2, 245-252, (1997)
[4] T.Muira, L.Davis, A Study of Ferrite Nonlinearity Evaluation for Higher Harmonic Generation Efficiency, *Proc.36th Eur. Microwave Conf.* 917-920 (2006)
[5] E.Schloemann, R.Bright, Broadband Stripline Circulators based on YIG and Li Ferrite Single Crystals, *IEEE MTT*, **34**, No12, 1394-1400 (1986)
[6] Y-S Wu, W. Ku, J. Erickson, A Study of Nonlinearities and Intermodulation Characteristics of 3 Port Distributed Circulators, *IEEE MTT*. **24**, No2, 69-77, (1976)
[7] E. Schloemann, Theory of Low Field Loss in Partially Magnetized Ferrites, *IEEE Trans. Mag*. **28**, No5, 3300-3302 (1992)
[8] D. Cruickshank, 1-2GHz Dielectric and Ferrites: Overview and Perspectives, *J.European Cer. Soc.*, **23**, 2721-2726 (2003)

Author Index

Author Index